Designer's Guide to the Dynamic Response of Structures

Designer's Guide to the Dynamic Response of Structures

Alan Jeary

City University of Hong Kong

E & FN SPON

An imprint of Thomson Professional

London · Weinheim · New York · Tokyo · Melbourne · Madras

Published by E & FN Spon, an imprint of Thomson Professional, 2–6 Boundary Row, London SE1 8HN, UK

Thomson Science & Professional, 2–6 Boundary Row, London SE1 8HN, UK

Thomson Science & Professional, Pappelallee 3, 69469 Weinheim, Germany

Thomson Science & Professional, 115 Fifth Avenue, New York, NY 10003, USA

Thomson Science & Professional, ITP-Japan, Kyowa Building, 3F, 2-2-1 Hirakawacho, Chiyoda-ku, Tokyo 102, Japan

Thomson Science & Professional, 102 Dodds Street, South Melbourne, Victoria 3205, Australia

Thomson Science & Professional, R. Seshadri, 32 Second Main Road, CIT East, Madras 600 035, India

First edition 1997

Typeset in 10/12pt Sabon by SPS, Hong Kong
Printed in Great Britain at The University Press, Cambridge

ISBN 0 419 18730 8

A catalogue record for this book is available from the British Library

♾ Printed on permanent acid-free text paper, manufactured in accordance with ANSI/NISO Z39.48-1992 and ANSI/NISO Z39.48-1984 (Permanence of Paper).

Contents

Preface

Much of this book has been written in the territory of Hong Kong, just prior to the change of sovereignty from Britain to the People's Republic of China. It is a time of great change, and this change has been occurring not just in the political arena. The changes to Hong Kong's built environment have been dramatic. During the observation of these changes over more than a decade, much new information has become available, and much of this appears here.

There are many references to tests conducted on full-scale structures in this book. I was fortunate to work with several groups of highly competent colleagues. I am indebted to these people for their energy and enthusiasm in helping me along the road of what has seemed like a lifetime quest, to try to understand just a little better how it is that structures really behave.

I would like to pay tribute to those people who have made the production of this book possible in these tumultuous times: firstly, to my secretary, Frances Wong, who has risen magnificently to all the seemingly impossible demands I have placed on her abilities and good nature; to my colleagues at the City University of Hong Kong, who have supported me through the turbulent times; and finally to my two sons, Ian and Thomas, who have made everything worth while.

Alan Jeary
Hong Kong
April 1997

CHAPTER 1

The dynamic performance of structures

It is perhaps rather unusual to begin a book with an invitation to the reader to discard the majority of what he or she may have been taught about the subject. In the case of the dynamic behaviour of structures, the invitation is based on the fact that most of what is conventionally written about the subject is dressed up in a mathematical mystique that is not really warranted, and serves only to maintain the subject in the hands of a select few who can handle the sometimes complex mathematics required.

There is an alternative. An approach to many problems, not just those of this particular subject, can be considerably simplified by using a different philosophical approach from the conventional. This approach is termed the 'systems approach' and is a subject area that is well developed in other disciplines. In particular, the electronics field has used a systems approach to some quite complex problems for years. The result is a simplification of the complex issues and the potential for a quantum leap in understanding by the mathematically urbane.

It is perhaps worth while to give an example here of the type of thinking that makes up the systems approach.

The often-quoted example of a refrigerator inside a locked room serves to illustrate the point. In this example, the refrigerator door is open and the question to be answered concerns whether the room temperature will rise or fall. You are effectively asked to estimate whether the cooling effect from the inside of the refrigerator will be greater than the heat dumped at the rear of the device. Looked at in these terms, it is difficult to assess which would be greater. With a systems approach, however, the solution becomes immediately obvious. In this example we merely draw a system boundary and look at what crosses the boundary. If we choose the boundary to be the edges of the room itself, then it quickly becomes apparent that crossing this boundary is only electricity, entering the room. The walls are assumed to be perfect insulators, and therefore the room must heat up. If we go on to assume that the walls are not perfect insulators, then this approach shows that the problem reduces simply to a calculation of the difference between the energy input in the electricity and the heat flow through the walls.

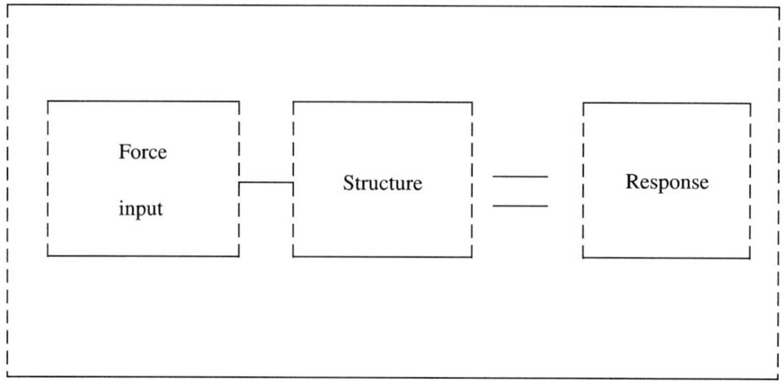

Fig. 1.1 Diagram of the systems approach to the assessment of the response of structures.

When the systems approach is applied to the assessment of the dynamic response of a structure, there are simply three elements to the calculation. These are represented in Fig. 1.1 and are the following:

1 Representation of the force input (F)
2 Representation of the structure (H)
3 Representation of the response (X)

It is common these days to use a representation of these parameters that is distributed according to frequency content, and so the mystique of the subject is built up. However, when this is put into a systems context, the rationale for all of today's methods of calculation becomes much clearer. The basic relationship involved merely says that, if you take the three representations above, and multiply F by H, the result is X, the response. Put simply this is

$$F \times H = X \tag{1.1}$$

The word 'representation', used above, is the key to the complexity that results from this very simple relationship. The representation could be in terms of frequency distribution, complex frequency response functions, matrix equations or even finite elements. The basic problem with understanding is that the equations used are rarely separated into this simple form so that a basic understanding can be obtained.

It is worth while spending a few moments to consider the implications of equation (1.1) above, for there are many.

(i) The entire structural representation is through a single parameter (H). The complexity of the structure can be represented in this form, and the significance is that the structure itself integrates all the forces acting upon it and behaves in a well defined way that can be described simply.

(ii) Equation (1.1) does not define whether the response is dynamic or static. In fact, the equation encompasses both types of behaviour, and, whilst this book deals mainly with the dynamic performance, equation (1.1) should always remind the reader that statics is merely a special case of dynamics in which the frequency involved is extremely slow. All equations presented here that are applicable to dynamics are capable of being reduced to statics equations merely by setting the frequency to zero. In this context, the generality of equation (1.1) can be appreciated.

(iii) Equation (1.1) puts together the force and the structure. If complex terms are used to represent both force and structure, then some complex interaction terms occur. In practice, these terms are the subject of great difficulties in modern codification. The reason for this is that it may not be immediately obvious exactly what causes a change in the response, whether in fact it comes from some characteristic of the force or of the structure itself. It is apparent in some codes of practice that confusion of this basic fact has occurred, and safety factors are placed on loading, for instance, when the uncertainty may well be concerned with the properties of the materials comprising the structure.

(iv) Equation (1.1) does not differentiate between different fields of study. The same equation holds good for the fields of wind, earthquake, ocean and highway engineering, and so on for all possible subjects that may occur. The calculation of response is basically the same problem independent of the source of the force.

(v) Equation (1.1) is a simple expression that can be rearranged in simple form, so that several other possibilities arise. For instance, if the response is divided by the representation of the structure, then the result is a representation of the force input. Alternatively, division of the response function by the force function will result in a representation of the structure. If the response and the force are measured in complex form, then the resulting detail about the structural behaviour will similarly be detailed.

For the assessment of the three elements of the basic equation (1.1), it is necessary to consider a number of issues, and here there are considerable differences in what is possible in the different fields. However, these differences are caused mainly by the statistical differences in the data sets rather than in any fundamental difference in the subjects. There is, of course, the added complication of non-linearities of the forces or of the structural behaviour. In practice, the structural non-linearities are by far the more complex, and this will be considered later. The approach to non-linearities generally is to take a view of the entire process at a time or a response level at which the non-linear property can be defined as though it were a linear equivalent for that circumstance only.

In general terms, the approach to the treatment of non-linearities is what separates such subjects as wind and earthquake engineering, rather than any fundamental differences in the underlying philosophy.

The handling of the characterization of the structure can also be made to be largely independent of the type of structure involved. The same techniques can be used in many different fields. There would, however, be a large difference in the assessment of the response. What would be acceptable for a highway, may be totally unacceptable for a residential building. The differences between highway and building performance would be apparent in the significance of the assessment of the response. A particular response or type of response of a bridge may be acceptable, whereas the same may not be true for a building with human inhabitants.

It is an interesting phenomenon that structures integrate all the effects of their structural performance. If viewed in systems terms, the uncertainties associated with such things as load sharing, edge conditions, variations of material properties throughout the structure, etc., are all averaged into the overall characterization of the structure. This characterization then represents the real behaviour of the real structure. This theme is central in this book. Instead of accumulating information about small parts of a structure, summing them together, adding up the effect and claiming that this represents the real behaviour, the approach taken here is to consider only the system of the entire structure itself. This information is distilled from real information about the complete behaviour of entire structural systems. In such measurements, all the uncertainties have been added together, summed and integrated to form a picture of the actual behaviour. Measurements of such integrated responses form the basis of the characterization of the structural systems that will be used throughout this text.

In this respect this book takes a 'top-down', or holistic, approach, rather than the 'bottom-up', or summation of the parts, philosophy that is commonly used. This basic difference in philosophy also simplifies the mathematics that is necessary to be able to deal with complex issues concerned with the dynamic response of structures. Since the approach derives from the real behaviour of real structures, it also represents a significant improvement in the precision with which structures can be assessed.

In subsequent chapters, attention will be turned to the representations of the force in different fields of study, of different types of structure, and of appropriate representations of the response of structures.

Information presented using this systems approach lends itself particularly to the design process. In Chapter 2, the context of the use of the systems approach for the design engineer is considered.

When users are well-versed in the use of the systems approach, they will find it an aid to understanding at an intuitive level, in much the same way that calculators made arithmetic more accessible than did the use of pencil and paper. The criticism that ease of use leads to abuse and misuse, coupled with a lack of understanding at a basic level, could apply to this subject as well if users do not continually bear in mind that the systems approach is merely a tool. The unthinking use of tools is dangerous, but their intelligent use allows quantum leaps of understanding and increased accessibility to an entire subject area. This is a plea that there is a germ of truth behind the Luddite philosophy that new technologies have dangers associated with them, but that there is an alternative to their methods of assuaging their fears through negation. This is an invitation to take the systems approach and to apply it intelligently to the subject area of the dynamic behaviour of structures.

In certain ways there are larger systems with which we must also be concerned. Human adaptations of the environment are now on such a large scale that entire cities form a system. In a case such as Hong Kong (Fig. 1.2), the city itself forms such a complex physical structure, positioned very close to the sea, that the interaction with the principal form of action from strong winds requires an added degree of sophistication to any analysis. In the particular case of Hong Kong, the situation is made even more complex by the close proximity of buildings, causing interaction, and excitation from traffic (Fig. 1.3) and from construction (Fig. 1.4). In a case such as San Francisco (Fig. 1.5), the earthquake risk is compounded by the use of infill in some areas. The juxtapositioning of sea, city and mountains is exemplified by the Lagoa region in Rio de Janeiro (Fig. 1.6), although in this case the wind action is rather less important than issues of economically induced quality assurance.

Fig. 1.2 Hong Kong, harbour area, 1997.

Fig. 1.3 Hong Kong Island, 1992.

Fig. 1.4 Hong Kong, construction close to residential dwellings.

Fig. 1.5 San Francisco, looking south.

Fig. 1.6 Lagoa region of Rio de Janeiro.

Fig. 1.7 Manhattan area, New York City, USA.

Our modifications to our planet invite us to consider matters in systems and systematic terms. Thinking in such terms presents the analyst with an advantage in considering matters on a larger scale, and with a view to consideration of all pertinent matters, even those which are not considered normally to be part of the problem. It is the author's intention that this book will provide the reader with the framework for such analyses, and for a method of thinking that will allow the consideration of all matters that are pertinent to a problem. The overall objective is to demonstrate that the analysis of structures has an underlying theme that is independent of the traditional themes such as those of wind or earthquake engineering. It is to be hoped that modern cities (Fig. 1.7) will be designed so that the risks associated with loading from any action will be reduced to an acceptable level, and that this book will contribute to that process.

CHAPTER 2 *Design requirements*

At this early stage, the requirements for design are dealt with in outline only. At a later stage, the detail will be considered.

At the design stage, there are several criteria that must be assessed. Designers must ask themselves precisely what are the criteria for design. A number of possible criteria are the following:

- Comfort of the occupants or users
- Survival of the structure
- Damage limited to areas that can be repaired

Clearly a choice here involves an assessment of risk. This risk involves the chances of a 'design magnitude force' (or larger) occurring and the chances of damage occurring when such an event takes place.

Designers must then assess what forces are likely to impinge on the structure. Whilst this will depend on the field of major concern (such as wind or earthquake), perhaps the most important division is into the categories of deterministic and random. These forces will be considered in turn.

2.1 *Deterministic forces*

Deterministic forces are those which can be defined mathematically. They can be represented by simple mathematical functions, and can be assumed to lend themselves to calculation quite readily. An example might be the forces induced by machinery operating. In such a circumstance, the force would be proportional to the unbalanced weights in the rotating parts of the machine, and such a force would simply exist when the machine was switched on.

2.2 *Random forces*

As far as the designer of a structure is concerned, random forces are mainly those forces which result from natural events such as wind storms or earthquakes. In practice, there are two main categories of 'random'. These are stationary random and non-stationary random. In all present-day codes of practice for wind or earthquake actions, it is assumed that random forces are stationary, at least for the purposes of analysis. This implies that the underlying physical mechanisms are unchanging with time. Clearly this cannot be the case. The wind comes from different directions and with gusts that are unpredictable. Earthquake activity may last for a few seconds or as long as minutes. Additionally, the ascribing of return periods for these events tends to suggest that they occur with clock-like regularity in an unchanging world. There is now plenty of evidence to suggest that we ourselves change many of the Earth's natural regimes. An example would be the mooted changes to the Earth's climate. In the main this is not a major concern for the designer. However, there are some very specific instances in which it is of considerable concern. This will be discussed later.

If, then, we assume a stationary random regime for weather or earthquake activity, we need to consider what the implications are. The requirements of the design engineer are to arrive at a characterization of F in equation (1.1) (see also Fig. 1.1) that will represent the force for which he or she wants to design. This value must be obtained first by assessing the governing design criterion from the three categories listed above. The designer must then decide what is the natural environment in the particular area in which his or her structure will be located. There are many sources for obtaining such force data.

Most of these are based on years of study of the occurrence of extreme events. The designer will often be confronted by a 'fifty-year wind' or a 'fifty-year earthquake'. The presentation of such random data in this form is misleading, since there is a suggestion that such events occur at the stroke of midnight as the calendar passes the 50 years mark. Clearly, stated in these terms, such an assumption is ridiculous, yet the impression remains that many involved in such assessments implicitly assume that statistics couched in such long-term phrases imply that there is no current possibility of such an event. Such assumptions are nonsense and are dangerous.

It is less misleading to term the 'fifty-year event' as a yearly probability of 0.02 of occurrence. This means that the annual probability of occurrence of such an extreme event is 0.02. This point is emphasized in Fig. 2.1. In this, a 5000-year history of the Earth is considered. During such a period, it would be expected that there would be 100 'probability 0.02' events, but, as can be seen, these occur randomly. There are sequences in which no 'fifty-year event' occurs for 150 years, and others in which up to four such events occur within the space of 50 years. It should be remembered that, despite such occurrences, the underlying statistic is unchanged.

It is interesting, though, to speculate on what might be the reaction of designers born at some stage during the sequence shown in Fig. 2.1. During a sequence in which there is no 'probability 0.02' event for an extended period, a designer could be excused for assuming that there may be some change to the environment. When this sequence is extended through the life

*	*	*	* *		*			**	*
	* *	*	*	* *		* *	*		**
***		*	***		*		* *	* *	*
			*	*	**		*		* *
*		* *	*			* *		**	
			*	*		*	**	*	**
*	*	* *	* *	**			*	***	
	*	*		* *	*	**			*
*	* *	*	* *		* *			* *	
* *	* *	*		**		* *		****	*

Fig. 2.1 A 5000-year sequence of 'fifty-year events'.

of his or her child and grandchildren (who also conveniently happen to be designers too!), it is probable that the local code-writing bodies will vote that there has been a change and that the requirements for design should be reduced. All this would occur within a stationary random sequence in which no change was occurring. The occurrence of a 'probability 0.02' event after this long sequence would then expose a deficiency in the code requirements, most probably through the medium of failures of structures or components of them.

It will be noticed that, in Fig. 2.1, 63 of the 5000-year sequences have 'fifty-year events' in them. This graphically illustrates the fact that there is an approximately 63% chance ($(1 - 0.02)^{50} = 0.364$), in any 50-year period, of the 'fifty-year event' occurring in that period.

The point of this is that the human species is impatient, whilst the time constants associated with natural processes are long. We arrogantly assume that, after having collected only a hundred years' worth of records, we can easily predict what will happen over the next several thousand. Clearly the confidence in statistical properties must be built up over a considerable period of time. We are only at the stage at which our model fits the data that we have, but which may need to be refined later. Refinement of statistical techniques allows us to refine these estimates of natural processes fairly quickly. A case in point is the Danish wind code, in which data for wind storms existed only for 10 years, yet, by using data for all severe storms that occurred within a year rather than the previously used maximum annual event only, the Danes were able to assemble good-quality wind statistics in a remarkably short period.

In the case of earthquake engineering, we are often dealing with much longer-period events. Within recent history, there have been totally unexpected earthquakes in Raleigh, North Carolina, USA, and Newcastle, Australia. These events merely point to the fact that the return period is very long, or that the probability of occurrence is very low. Nevertheless, an event must occur at some time, and the fact that it occurs now does not alter the fact that they are very low-probability events. The changing of earthquake risk maps to reflect these occurrences merely serves to amplify the fact that we are drawing such maps before all the data that we really need are available.

This consideration of random forces is one of the most important themes of this book. Since the force input to the structure is one of the three main elements of the systems approach, then quite clearly arriving at a useful description of the force is of paramount importance. Deciding whether the force is random or deterministic is a fundamental choice. For the majority of applications, however (and especially when naturally occurring forces are involved), then the force is likely to be random. Given that a force will be random, then a choice must be made about the nature of the statistics involved. These will be governed essentially by the physical location being considered, the physical process, and then any non-linearities that may be significant.

2.3 *Characterization of the structure*

At first sight, it may seem odd that it can be asserted that the characterization of all structures can be performed in a similar manner. In practice, especially when considering dynamic behaviour, this is a truism, and is another consequence of using a systems approach to the problem. The way that this is achieved is to consider the overall behaviour of the entire structure.

The complex response can be built up with a series of deflected shapes, each of which corresponds with what is termed 'a natural mode of vibration', or 'mode shape'. The sum of all these mode shapes is the total deflected shape of the structure under the loads to which it is subjected. Since the mode shapes are specific to the individual structure, then the summed response is also unique to that structure.

Each type of structure has a series of mode shapes associated with it, and the components of each mode shape are specific to that individual structure. Figure 2.2 shows idealized mode shapes of the first three translational modes of tall buildings, chimneys and bridges.

In this approach, the total response can be built up by adding together the components from each individual mode of vibration. The total response of the structure is a combination of the shapes summed in the proportion appropriate to the amount of energy dissipated in that mode.

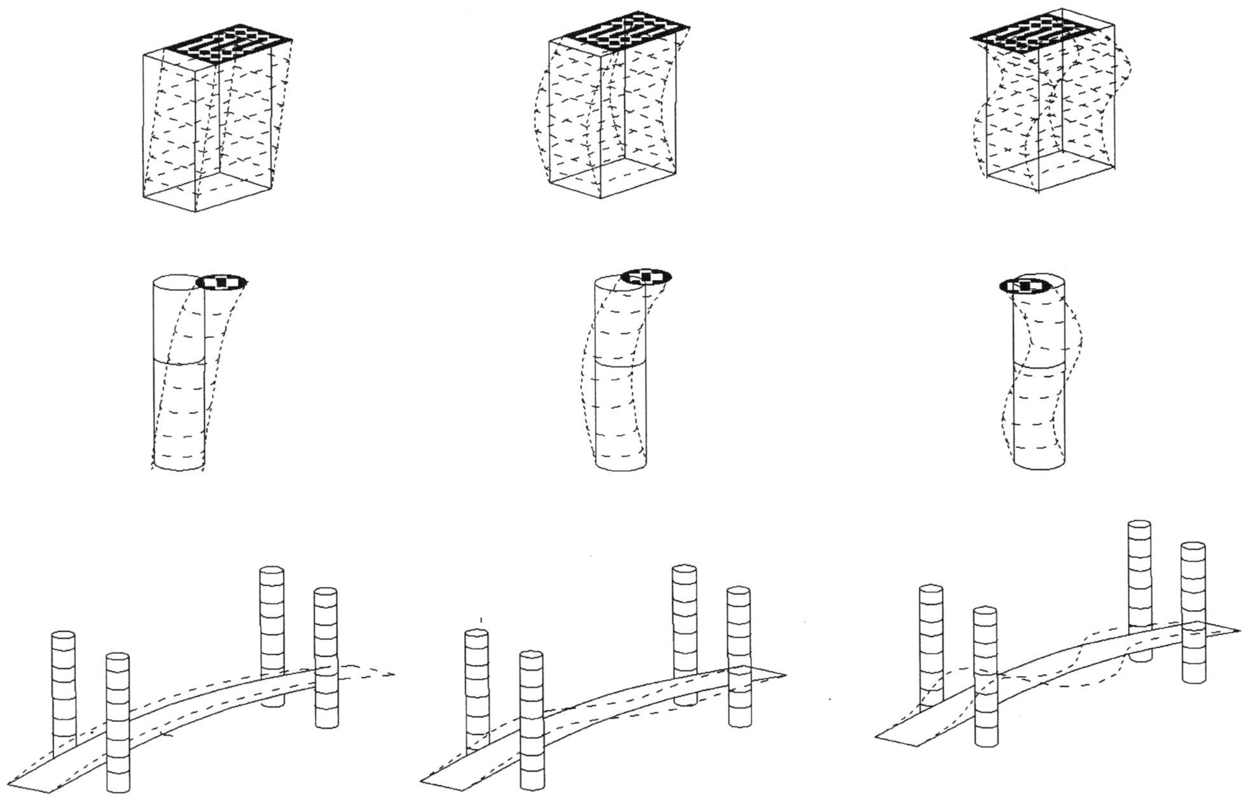

Fig. 2.2 Idealized mode shapes for some common structures.

There is an alternative method for the representation of each mode of vibration. This is termed the 'frequency response function' approach. In practice, frequency response functions lend themselves particularly well to the systems approach. They are considered in detail in the next chapter. Both of the approaches mentioned above rely on the principle termed 'modal superposition'. Modal superposition normally implies that no non-linearities are present and that the overall response will be the simple geometric summation of the responses in the individual modes of vibration.

The forces acting on structures as a result of different types of action (such as wind, wave, traffic or earthquake) are considered in more detail in the next chapter. The most important consideration for the present is that they can all be represented as being one of the three elements of the systems approach (the F function) and they can all be represented by similar means.

2.4 *A philosophical approach to design*

The question of conduct, and a type of 'Hippocratic oath' for engineers, naturally reinforces the fact that design is heuristic in nature. Koen (1984) described five types of heuristic commonly applied to design:

- Rules of thumb
- Factors of safety (sometimes called factors of ignorance)
- Social attitudes
- Risk management
- Resource allocation economics

Interestingly, Koen went on to argue that the design process involves so much iteration and feedback that it should be regarded as a guideline rather than as a rule. (This factor is highlighted by the current confusion in the UK over differences between British 'Standards' and 'Codes of practice'.) He suggested that design is a morphology, a set of heuristics useful for solving an engineering problem. He summarized the engineering method as being the use of engineering heuristics, including all those indicated, with the purpose of causing the best change in a poorly understood situation within the available resources.

This attempt to formalize an understanding of engineering, engineering practice and the role of the engineer in society, is itself a branch of engineering, and is typified by the use of analogies. Professional engineering bodies or councils themselves have recourse to reference to terms that are more normally associated with the arts. Witness is given by the following definition of engineering by the Engineers' Council for Professional Development in Australia: '...the profession in which the mathematical and physical sciences gained by study, experience and practice is applied with judgement to develop ways to utilize economics, materials and forces of Nature for the progressive well-being of mankind'. The emphasis on judgement, testing of decisions, economics and progression of human aspirations clearly removes engineering from science, and just as clearly shows why the present distrust of the engineer in society has developed. (The interesting siting of economics first in the list is presumably a currently fashionable anomaly.) The engineer is quite clearly not trained to perform even one-half of the duties implicit in this definition. He or she is expected to acquire such a feel for the well-being of mankind, and judgement, through some mysterious experience-gathering exercise. The difference between this theory and present-day practice is profound.

2.5 Design in a changing world

A timely occurrence for the purposes of this discussion is the fairly recent Northridge earthquake (near Santa Cruz, in northern California). The Nimmitz Highway in Oakland was designed to the limits of knowledge 30 years ago. Many similar structures in that area were strengthened in the interim, as more knowledge about earthquake forces became available. The failure of the double-deck section of the highway in the 18 October 1989 earthquake was a social failure as much as an engineering one. The highway engineers responsible for the continuing use needed to evaluate the current state of knowledge and consider this in the light of the regulations in force at the time of construction. Clearly this process failed. The question to be asked is: 'Was this a failure of engineering, engineers or education of engineers?' Undoubtedly, engineers will bear the responsibility, but, if they were not prepared in their education to keep abreast of the most modern developments, is this just? In some ways, this book is a reaction to these events, as it attempts to make available knowledge that is not generally taught at undergraduate level.

One possible scenario for the design process is the following:

1 Review of a problem
2 Definition of requirements
3 Synthesis of possible solutions
4 Modelling of alternatives
5 Analysis of alternatives
6 Selection and reporting of preferred solution
7 Implementation

There is feedback and iteration at all of the seven stages, and the approach to be followed should be largely heuristic.

2.6 Design considerations

The 'serviceability limit state' (SLS), since it occurs at lower loads than the 'ultimate limit state' (ULS), will normally dictate the design. However, the consideration of exactly what constitutes a serviceability requirement is often changed by the results of new information about the behaviour of buildings and structures. For instance, only 15 years ago it was considered that structural loads being applied to non-loadbearing parts of a structure would constitute a limit state of serviceability. Today it is recognized that a great deal of load sharing occurs in practice. The threshold of perception of motion often forms the real practical lowest load serviceability criterion.

Corrosion, from the effects of chloride attack, is probably also a prime reason for failure to meet a serviceability criterion in Hong Kong.

2.7 The codification process

Codes of practice attempt to force designers to exploit all means available so as to maximize safety, durability and economy. Codes of practice are themselves, of necessity, transient in nature, and merely represent the best advice available at the time of drafting. This suggests two things. Firstly, the quality of codes of practice is a function of the team drawing them up; and secondly, their usefulness is a function of the information that has become available since the formalizing process of giving the code of practice a legal status. In most jurisdictions, a code of practice represents the minimum standards that are acceptable. A designer is free to go beyond the scope of a code of practice if, in so doing, he or she increases the safety, efficiency or performance of his or her structure.

There have been moves recently towards the use of 'limit states'. Such limit states represent the threshold between acceptable and unacceptable behaviour. However, this process is not complete, and we currently have a mixture of design principles.

2.8 Code formats

These can be classified in the following ways:

1 Philosophy – deterministic or probabilistic
2 Methods and accuracy of calculations
   (a) Working stress design (WSD)
   (b) Ultimate stress design (USD)
   (c) Limit state design (LSD)

The LSD has coincided with the introduction of probabilistic methods. However 2(a) and 2(b) above have absorbed some probabilistic methods.

2.8.1 Working stress design (WSD)

This method was adopted during the first half of the nineteenth century, and reflected humankind's confidence in prevailing over all things natural.

The basic idea is to impose an upper limit to the stresses acting on all parts of the structure simultaneously, so that materials remain in the elastic range. A 'factor of safety' is then included. The theory of elasticity can then be applied. Upper limits for stresses caused by various loads (such as bending and shear) under service conditions are then calculated, and a factor of safety (normally 2 for steel and 3 for concrete) is applied. This factor of safety is normally on the strength of the materials, with the result that they operate in the elastic range. Margins of safety with respect to failure are therefore at least 2. However, concrete is assumed not to have any tensile resistance, and this leads to a risk of cracking.

Doubts about this approach arose when actual margins of safety were evaluated, and it was concluded that:

1 Linear elastic behaviour does not always occur in practice under normal operating conditions.
2 Calculations of response based on linear elastic theory do not always give a good indication of immediate precursors to failure. If there are load combinations, then the linear elastic approach may not be a good indicator of combinations.
3 Engineering rules about material behaviour are based on an average of a large number of very small effects. When effects such as cracking or creep are considered, then there is no alternative but to use non-linear approaches to modelling.
4 The interaction of elements of a structure cannot be modelled. This is why the experience that is usable by a code formulation committee is of paramount importance. For this reason, factors of safety are applied. A major problem is that these are made greater when more uncertainty about real material behaviour is encountered. Such factors have alternatively been described as 'factors of ignorance'.

There are many complex situations that are not modelled in such circumstances. For instance, the residual strength of cracked sections cannot be considered, and load sharing by so-called 'non-structural' elements is simply ignored. Additionally, a structure's ability to redistribute load into alternative paths after the initiation of localized damage is not considered at all.

2.8.2 Ultimate stress design (USD)

This is also known as the 'load factor approach'. This involves taking the failure state, and establishing a ratio that must exist between the ultimate load and the service load. Such an approach can also be combined with a probabilistic relationship.

2.9 Definition of limit states

A structure or part of a structure is rendered unfit for use when it reaches a particular state known as a 'limit state', in which it ceases to fulfil the function or satisfy the demands for which it was designed. The limit states can be placed into two categories – ultimate limit states and serviceability limit states. The former normally relates to the load-carrying capacity of the structure, whereas the latter refers to normal use or durability.

In a structure, a limit state may be reached as the result of a number of random factors, which combine and affect the safety factors. These lead to the following:

- Uncertainty about the values used for strengths of materials.
- Uncertainty about the assumed geometry of the structure or sections of it.
- Uncertainty about permanent loads, variable imposed loads, or imposed deformations.
- Departure of the actual loading effects from the calculated values.

The objective of the limit state approach is as follows: the probability of reaching a particular limit state must be kept to a predefined value.

There are two categories of limit state: ultimate limit state (ULS) and serviceability limit state (SLS).

2.9.1 Ultimate limit state (ULS)

The main subdivisions for this are listed below:

- Loss of equilibrium of a part or the whole of the structure, when considered as a rigid body.
- Rupture of critical sections or parts of the structure.
- Transformation of the structure or parts of the structure into a mechanism.
- Instability by deformation.
- Deterioration arising from fatigue effects.
- A change of geometry of a structure, caused by cracking or deformation, that requires repair.
- An ultimate limit state may be the result of accidental loading (such as explosion, fire, collision or overpressure).

2.9.2 Serviceability limit state (SLS)

The main subdivisions for this are the following:

- Human perception of motion.
- Excessive deformations.

- Cracking, leading to excessive maintenance.
- Deterioration and corrosion, especially that causing surface staining.
- Differential settlement of parts of the foundations.

In considering the dynamic performance of structures, the SLS becomes particularly important. Human response to vibration is often a limiting factor in the use of a structure (where humans are users), and the irrational generalized comfort criterion has often been stated to be that the level of response must not be perceptible. Unfortunately, people have different perception thresholds, and different parts of a structure may be more prone to vibration. The siting of directors' suites at the top of tall buildings is a particularly unfortunate design limitation in this respect.

The approach in this book is intended to allow the designer sufficient flexibility to work with any of the above-mentioned formats. However, over and above this, it is intended that the systems thinking that it induces will instil in the designer an ability to judge when codes of practice should be used with caution.

CHAPTER 3

Characterization of forces and structures by frequency response functions

The purpose of this chapter is to invite the reader familiar with the conventionally used concepts of stiffness, load and deflection to alter perspective slightly and to use the input–system–response concept. When stated in this manner, it can be appreciated that there will be little difficulty if the system and the stiffness were to be the same. What occurs in practice is that the stiffness method is merely a special case of the more general systems approach.

It is also true that the systems approach, being general, has other special cases. For instance, if frequencies are all set to zero, what results is the conventional quasi-static stiffness–load–deflection equations.

There is a great advantage to be gained by using what are termed 'frequency response functions' for the characterization of not only structures, but also forces and deflections. In this way, a simplified and unified systems approach is introduced.

3.1 Frequency response functions

It is first necessary to define a frequency response function. In doing this, we will avoid the mathematical complexities, and merely refer to the part that will be useful for structural analysis.

In arriving at a definition of a frequency response function, we must first define the transfer function of a system in terms of a Laplace transform:

$$F(S) \cdot H(S) = R(S) \tag{3.1}$$

The Laplace operator is a complex function, and Appendix F gives some details, definitions and some of the more common transforms. In general terms, the Laplace operator can be used to transform differential equations to algebraic ones.

For our purposes, we reduce this transfer function relationship (3.1) to a complex frequency relationship by letting the operator S tend to the imaginary form only. Thus

$$S \rightarrow \mathrm{j}\omega$$

where j is the square root of -1, and ω is the circular natural frequency. It should also be noted that $\omega = 2\pi f$. This contrivance allows the frequency response function relationship:

$$F(j\omega) \cdot H(j\omega) = X(j\omega) \tag{3.2}$$

It is apparent that the 'j's here carry information about the phase relationship between the input and the output. It is also interesting that, whilst transfer functions can strictly only be used as a mathematical function expressed using Laplace operators, the frequency response function lends itself to expression in either mathematical or graphical terms. It is the graphical representation that is the more useful, and perhaps the more understandable for practical situations.

Equation (3.2) is basically all there is to most modern codes of practice dealing with the dynamics of structures. The frequency response function for the force is defined; the function $H(j\omega)$ is a representation of the structure built up through the use of information about each mode of vibration; and the product results in the displacement (or some derivative from displacement) as a function of frequency. That this representation can be presented graphically is perhaps the single most significant factor in the successful implementation of the modern methodology.

In simple terms, equation (3.2) can be represented graphically as in Fig. 3.1. The simplicity of this can now be appreciated. The response (X_1) is merely the product of the force F_1 and the representation of the structure $H(j\omega)$. This representation of the structure is actually a further simplification in that the modulus of the term allows a single graph to represent the structure, rather than the two that would be necessary if the phase information were also included.

In practice, the three examples in Fig. 3.1 actually demonstrate quite different situations. F_2 is typical of a long-term naturally occurring force such as that generated by wind. The corresponding response can be seen to be distributed against frequency in the same general manner, but with the characteristics of the structure superimposed.

F_3 is typical of a single-frequency source such as machinery. Since there is very little energy at other frequencies, then the response shows a similar form, and the response is also single-frequency. If the frequency of the force is changed, as would occur when the speed of rotating machinery changes, then the frequency of the input force will change. The response is then modified by the frequency response function of the structure, and the resulting response will be greater or lesser depending on where in the structure's characteristics the force occurs. An inspection of the structure's frequency response function shows that at certain frequencies the response will be multiplied by a factor that is much larger. These frequencies correspond to the resonance conditions of the structure, each one also having its characteristic deflected mode shape.

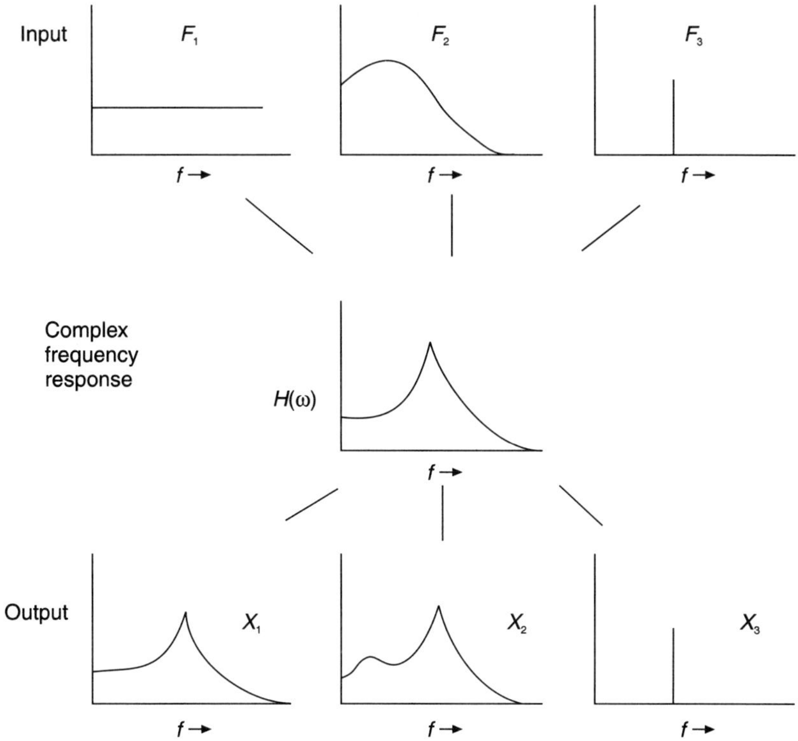

Fig. 3.1 The spectral approach: input × complex frequency response = output.

The changing of an input frequency until it coincides with a resonance of the structure is rather similar to the process of tuning a radio or television to a broadcasting station. When the frequency is correct, then the response is amplified. In the structure, this resonance is characterized by the dynamic magnification factor, and different values of magnification will occur at each of the resonance conditions of the structure.

The force represented by F_1 in Fig. 3.1 is an artificial condition in which the force function is termed 'white noise'. In this case the response is merely the constant value (against frequency) of the input multiplied by the frequency response function of the structure. This condition is demonstrated to amplify the point that the response reflects the structure and the forcing function. All response functions of structures contain information about the force and the structure itself.

3.2 System mathematics for frequency response functions

Having seen the overall use of the frequency response function, it is now necessary to look at the graphical representation in a little more detail.

It is convenient to start with basics in arriving at the form of the mathematical representation of frequency response functions. The end-point is to show the elements of the frequency response function that correspond to the conventional equations normally associated with structural behaviour.

3.2.1 General derivation

There are several possible starting points for the consideration of the dynamic behaviour. However, in the ultimate analysis, all approaches are simply a mathematical representation of an energy balance. In the present instance we will start with Newton's second law of motion:

$$F = M\frac{d^2x}{dt^2} \tag{3.3}$$

For the analysis, we need to sum all the components, i.e. all the forces that create an acceleration of the mass of the structure. These forces comprise an external force $f(t)$, an elastic spring force f_s and a damping force f_c. Thus

$$F = f(t) + f_c + f_s \tag{3.4}$$

The elastic force can be described as

$$f_s = -Kx \tag{3.5}$$

where K is the spring constant and the force opposes the motion. The damping force is often described as

$$f_c = -C\dot{x} \tag{3.6}$$

where the damping force is assumed to be proportional to velocity and is termed 'viscous damping'. There is a full discussion about this assumption in Chapter 7.

Thus the equation becomes

$$f(t) - Kx - C\dot{x} = M\ddot{x} \tag{3.7}$$

or

$$f(t) = M\ddot{x} + Kx + C\dot{x} \tag{3.8}$$

This equation is normally used for all representations of the dynamic response of structures.

Noting that the Laplace operator can be used as a differential operator, we have the following:

$$(S^2M + SC + K)X = F(t) \tag{3.9}$$

It is convenient to solve this differential equation assuming merely that the system is perturbed by a unit step function. There are three possible types of solution depending on the value of the complementary function:

$$\frac{C^2}{4M^2} = \frac{K}{M} \tag{3.10}$$

These are:

- Overdamped
- Critically damped
- Underdamped

In practice, all structural systems are underdamped, although the concept of critical damping is a useful reference point. The complementary function derived from the solution to equation (3.9) is

$$-\frac{C}{2M} \pm \sqrt{\left(\frac{C^2}{4M^2} - \frac{K}{M}\right)} \tag{3.11}$$

and so the critically damped case occurs when the part underneath the square root sign is zero, i.e.

$$C_c = 2\sqrt{KM} \tag{3.12}$$

This defines a critical damping value. The actual damping is normally expressed as either a fraction or a percentage of this value.

If we now consider the underdamped case, we can solve the complete response of the system to a simple step function. It takes the following form:

$$x_0 = x\left\{1 - \frac{\exp[-(C/2M)t]}{\sqrt{1 - C^2/4MK}} \sin\left[\left(\sqrt{\frac{K}{M} - \frac{C^2}{4M^2}}\right)t + \cos^{-1}\left(\frac{C}{2\sqrt{KM}}\right)\right]\right\} \tag{3.13}$$

This daunting equation can be simplified and put into more readily usable engineering terms by dint of some simple considerations.

If we suppose that the damping in the system were zero, then the system would oscillate as follows:

$$x_0 = x\left[1 - \sin\left(\sqrt{\frac{K}{M}}\,t\right)\right] \tag{3.14}$$

This defines an undamped natural frequency in which

$$\omega_n = \sqrt{K/M} \tag{3.15}$$

If we use the definition of damping as a ratio of actual damping to critical damping:

$$\zeta = \frac{C}{2\sqrt{KM}} \tag{3.16}$$

then equation (3.13) can be rewritten as

$$x_0 = x_i\left(1 - \frac{\exp(-\zeta\omega_n t)}{\sqrt{1 - \zeta^2}} \sin\left(\omega_n\sqrt{1 - \zeta^2}\, t + \cos^{-1}\zeta\right)\right) \tag{3.17}$$

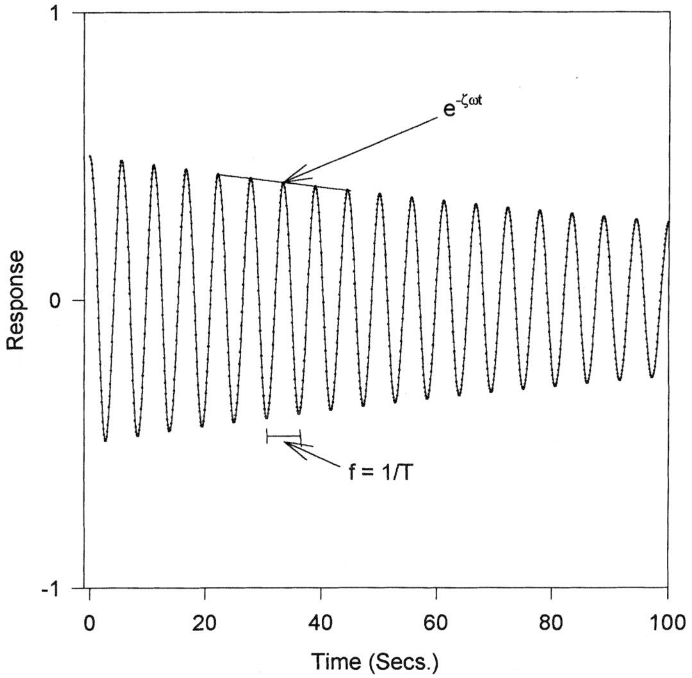

Fig. 3.2 Decay of oscillation: natural frequency = 0.2 Hz; damping = 0.5%.

and this defines an oscillation with an envelope of

$$x = \exp(-\zeta\omega_n t) \tag{3.18}$$

This can be represented graphically as shown in Fig. 3.2.

The envelope of these responses can be alternatively expressed as

$$x = e^{-t/\tau} \tag{3.19}$$

where τ is the time constant of the system and

$$\tau = 1/\zeta\omega_n \tag{3.20}$$

Time constants can occasionally be useful in the consideration of a system's response. A system, once given a perturbation, will return to its original state in a time that is described neatly by the use of time constants. To get within a certain percentage of the steady state, the time taken is as shown in Table 3.1. For practical purposes, then, it can normally be assumed that a system has reached close to steady state in four time constants, where the time constant is given by equation (3.20).

The equations all represent the response of a single mode or single degree of freedom of the structure. It is now possible to see that the response of a single mode of vibration can be described very simply. Equation (3.8) is

Table 3.1 Time taken to get within a certain percentage of steady state

Percentage	*Number of time constants*
63.2	1
86.5	2
95.0	3
98.2	4
99.3	5

perhaps the basis for most of the solutions to dynamics problems currently available, and indeed it is used in codes of practice.

The significance of this equation is now pursued somewhat to show the implications of its use and to show the similarities with conventional approaches to the design of structures. Equation (3.8) can be rewritten using the expressions for damping and frequency in equations (3.15) and (3.16):

$$\ddot{x} + 2\zeta\omega_n\dot{x} + \omega_n^2 = F(t)/M \tag{3.21}$$

This equation is general, but it can be made specific to a single mode of vibration or to a 'single-degree-of-freedom system':

$$\ddot{x}_r + 4\pi\zeta_r f_r \dot{x}_r + 4\pi^2 f_r^2 x_r = F_r/M_r \tag{3.22}$$

The 'r' subscripts refer to a mode labelled 'r', which has modal properties of frequency, damping, displacement (x_r) and mass. The concept of a modal mass, which is different from the total mass of the building, will be discussed later. Essentially, though, all of the parameters with the 'r' subscript represent quantities participating in the particular mode of vibration.

If we now return to the use of the Laplace operator S, equation (3.22) then becomes

$$(S^2 + 4\pi\zeta_r f_r S + 4\pi^2 f_r^2)x_r = F_r/M_r \tag{3.23}$$

We can now make use of the fact that the Laplace operator is composed of a real and an imaginary part. For real systems, it is necessary only to consider the imaginary part of the Laplace operator, and so equation (3.23) becomes

$$[(j\omega)^2 + 4\pi\zeta_r f_r j\omega + 4\pi^2 f_r^2]x_r = F_r/M_r \tag{3.24}$$

or

$$x_r = F_r\left(\frac{1/K_r}{1 - \frac{f^2}{f_r^2} + j2\zeta_r(f/f_r)}\right) \tag{3.25}$$

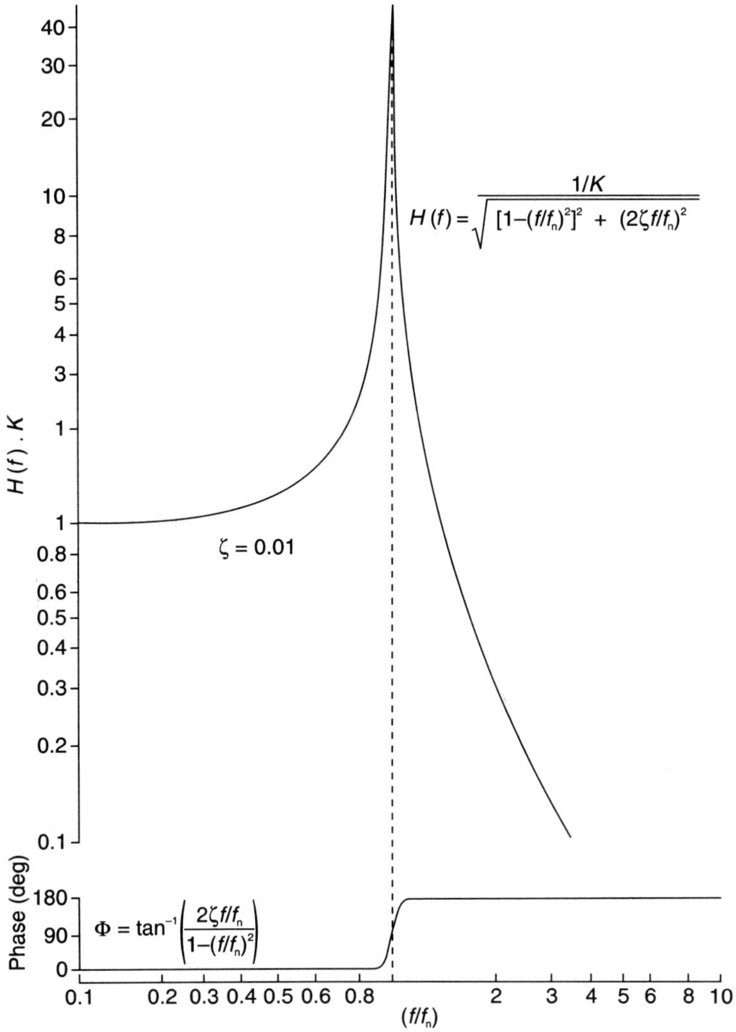

Fig. 3.3 Complex frequency response of single-degree-of-freedom system.

It is interesting to compare this equation with equation (3.2) and with Fig. 3.1. It can be seen that in equation (3.25) we have the force action (F_r) and the response (x_r) as before, but now the representation of the structure becomes the remaining part of equation (3.25). This part is called the 'complex frequency response function' ($H(f)$) of the structure. For a single mode with a damping ratio of 1% of the critical value, the complex frequency response function is shown in Fig. 3.3. As in this figure, the response is often depicted in two parts: the modulus and the phase of the response with respect to the force.

It is possible to investigate the form of equation (3.25) under different conditions. In particular, the resonance condition and the quasi-static (zero-frequency) conditions will be considered.

3.2.2 Quasi-static response

If we set $f = 0$ in equation (3.25), we obtain the following:

$$x(0) = F(0)/K(0) \tag{3.26}$$

This is the type of equation that underwrites the study of the quasi-static response of structures and says that the response can be calculated from the product of an applied force and the inverse of the stiffness.

3.2.3 Resonance response

If we set $f = f_r$ in equation (3.25), we arrive at the following:

$$x_r = F_r \frac{1/K_r}{j2\zeta_r} = \frac{F_r}{j2K_r\zeta_r} \tag{3.27}$$

This equation shows that the response at resonance is larger, by a factor of $1/(2\zeta)$, than the quasi-static response. Accordingly this factor, $1/(2\zeta)$, is termed the 'dynamic magnification factor'. The presence of 'j' in the denominator signifies that the response is shifted 90° with respect to the phase of the force.

Alternatively, if we substitute the relationship of equation (3.15) we obtain:

$$M_r = \frac{F_r}{8\zeta_r \pi^2 f_r^2 x_r} \tag{3.28}$$

This interesting variation shows that, if we have measured four of the five parameters (force, damping, frequency, response, mass), then we may calculate the fifth one.

This observation leads to some interesting possibilities. Firstly, a structure can be 'weighed'. Secondly, a calibrated structure (one for which the mass is known) may be used as a transducer to measure the force acting on it. Lastly, it is interesting that damping, frequency and response are actually non-linear parameters, but of course the mass of the structure remains constant. This means that the effect of all non-linearities should counterbalance when calculating the mass. In this way, a measure of extra confidence in the assessment of non-linearities can be added.

The systems dynamics approach can therefore be seen to encompass the condition with which most structural analysts are familiar. However, it extends to the more general condition of non-zero frequency.

Of course, the conventional approach is really unjustified, since, if the response of a system were really to occur at zero frequency, then no motion at all could occur, since even the most minute and slowest of movement must also require a small but finite frequency to be associated with it.

3.3 Modal parameters

It can be seen from the foregoing that the systems approach forces a relationship between the structural system, the force input to it and its response to that force. The systems approach makes no distinctions about the source of the force. Accordingly, it is a unifying approach to such diverse fields as wind, ocean, traffic and earthquake engineering. The only difference between these fields is the method of arriving at a description of the force, and of deciding whether the inherent non-linearities are of sufficient importance to merit a special approach. Needless to say, requirements for simplification inevitably lead to an approach that is essentially similar in all these fields. We merely approximate parameters at the different amplitudes associated with the particular field of study.

This is not to say that there are not significant differences in approach in the different fields. However, these are normally associated with deciding on an approach to the characterization of the forces involved, and of arranging an approach to the non-linearities. In some fields this is much more important than in others, with earthquake engineering being of particular importance here. Much of earthquake engineering practice is concerned with the assessment of the performance of individual elements within a structure that will be forced to undergo motion or relative motion beyond the limits of linear elastic theory.

To the practitioner who has not been brought up on dynamics, perhaps one of the more difficult concepts to grasp is that of modal parameters. Why should it be that a modal mass is different from the mass of the structure, or that a modal stiffness should be different from the conventional and well understood stiffness of the entire structure.

The reason for these differences is straightforward, and can be thought of in the following terms. If we consider stiffness, then there is a stiffness associated with each resonance mode of vibration of the structure. The overall stiffness of the structure is composed of the sum of the stiffnesses in all of the individual modes of vibration that would be excited by some forcing function. If we consider a dynamically simple structure, like a tall building, then the contribution from each mode of vibration to the overall stiffness of the building varies considerably for each mode of vibration. In practice, since more than 98% of energy dissipation normally occurs by first-mode bending, then the stiffness of this single fundamental mode of vibration approximates quite closely to the conventional static stiffness of conventional theory. The remaining 2% contribution comes from all other modes of vibration summed together. The contribution of each mode of vibration to the overall effect is governed by an influence function.

However, not all structures are as simple (dynamically) as a tall building. Nevertheless, if the influence function is considered, then the overall effect is the sum of the influences of all of the modes of vibration. For each type of structure, the number of modes of vibration that are necessary to consider, so as to get a good approximation to the overall effect, can be assessed from a consideration of the influence functions for each of the individual modes of vibration.

There also happen to be some fairly powerful mathematical tools that emerge as a result of an analysis of such influences. Whilst the mathematical introduction is not simple, the result is, and can be reduced to geometric proportions to allow dynamics techniques to be used for (for instance) load testing.

What follows is one mathematical treatment to allow the comparison of modal parameters.

3.3.1 Lagrange's equation

Lagrange's equation is merely an energy-balance equation. All energy in a system is considered and balanced. For our purposes we need to consider a dynamic generalized system in which there are j masses, S springs, W dampers and N modes of vibration. The types of energy we then consider are the following:

- Kinetic energy T
- Strain energy U
- Work done by damping forces W_c
- Work done by external forces W_e

Lagrange's equation can then be stated as:

$$\frac{\mathrm{d}}{\mathrm{d}t}\left(\frac{\partial T}{\partial \dot{a}_i}\right) - \frac{\partial T}{\partial a_i} + \frac{\partial U}{\partial a_i} - \frac{\partial W_c}{\partial a_i} = \frac{\partial W_e}{\partial a_i} \tag{3.29}$$

Expressions can then be derived for these four terms as follows:

Kinetic energy T

This is given by

$$T = \sum_{r=1}^{j} \frac{1}{2} M_r \left(\sum_{n=1}^{N} \dot{a}_{rn} \right)^2 \tag{3.30}$$

where $\dot{a}_{rn}$ is the velocity component of mass r associated with the nth mode of vibration.

Strain energy U

This is written as

$$U = \sum_{g=1}^{S} \frac{1}{2} K_g \left(\sum_{n=1}^{N} \Delta_{gn} \right)^2 \tag{3.31}$$

where Δ_{gn} is the distortion of the spring g in the nth mode of vibration and K_g is the stiffness of that spring.

Work done by damping forces W_c

Here we assume that the damping force is proportional to velocity, i.e. that it is viscous damping. Then

$$W_c = \sum_{v=1}^{W} C_{vn} \sum_{n=1}^{N} \dot{a}_{vn} \sum_{n=1}^{N} a_{vn} \tag{3.32}$$

where C_{vn} is the damping constant of dashpot v in the nth mode of vibration.

Work done by external forces W_e

Here we have

$$W_e = \sum_{r=1}^{j} F_r \sum_{n=1}^{N} a_{rn} \tag{3.33}$$

where F_r are the external forces.

A small normalizing transformation is convenient at this point. We assume that there is a displacement A_n of an arbitrarily chosen mass, and we rewrite the equations in terms of this variable. This normalization is the key to the use of mode shapes. It is convenient to normalize to the maximum displacement in any mode of vibration, although other forms of normalization are possible, and indeed are sometimes used. Thus we have:

$$a_{rn} = A_{rn} \frac{a_{rn}}{A_n} = A_n \phi_{rn} \tag{3.34}$$

$$\dot{a}_{rn} = \dot{A}_{rn} \frac{\dot{a}_{rn}}{\dot{A}_n} = \dot{A}_n \phi_{rn} \tag{3.35}$$

$$\Delta_{gn} = A_n \frac{a_{gn}}{A_n} = A_n \phi_{gn} \tag{3.36}$$

$$C_{vn} = A_n \frac{C_{vn}}{A_n} = A_n \phi_{Cvn} \tag{3.37}$$

$$C_{vn} = \dot{A}_n \frac{C_{vn}}{A_n} = \dot{A}_n \phi_C \tag{3.38}$$

Additionally, it is assumed that the modes of vibration are normal modes and that the principle of orthogonality applies. This implies that

$$\left(\sum_{n=1}^{N} \dot{A}_{rn}\right)^2 \rightarrow \sum_{n=1}^{N} \dot{A}_{rn}^2 \tag{3.39}$$

$$\left(\sum_{n=1}^{N} \Delta_{rn}\right)^2 \rightarrow \sum_{n=1}^{N} \Delta_{gn}^2 \tag{3.40}$$

Thus equations (3.30) to (3.33) can be rewritten as

$$T = \sum_{r=1}^{j} \frac{1}{2} M_r \sum_{n=1}^{N} \dot{A}_n^2 \phi_{rn}^2 \tag{3.41}$$

$$U = \sum_{g=1}^{S} \frac{1}{2} K_g \sum_{n=1}^{N} A_n^2 \phi_{\Delta gn}^2 \tag{3.42}$$

$$W_c = \sum_{v=1}^{W} C_{vn} \sum_{n=1}^{N} \dot{A}_n \phi_{Cvn} \sum_{n=1}^{N} A_n \phi_{Cvn} \tag{3.43}$$

$$W_e = \sum_{r=1}^{j} F_r \sum_{n=1}^{N} A_n \phi_{rn} \tag{3.44}$$

From the relationships in equations (3.41) through (3.44), it is possible to derive the relationships needed in Lagrange's equation (equation (3.29)):

$$\frac{\partial T}{\partial \dot{A}_n} = \sum_{r=1}^{j} M_r \dot{A}_n \phi_{rn}^2 \tag{3.45}$$

$$\frac{\partial}{\partial t}\left(\frac{\partial T}{\partial \dot{A}_n}\right) = \ddot{A}_n \sum_{r=1}^{j} M_r \phi_{rn}^2 \tag{3.46}$$

$$\frac{\partial U}{\partial A_n} = A_n \sum_{g=1}^{S} K_g \phi_{\Delta gn}^2 \tag{3.47}$$

$$\frac{\partial W_e}{\partial A_n} = \sum_{r=1}^{j} F_r \phi_{rn} \tag{3.48}$$

Additionally, it is assumed that

$$\partial T / \partial A_n = 0 \tag{3.49}$$

For the vast majority of structural dynamics problems, this is no problem. The only exception would be if the structure were losing mass at a significant rate, rather in the way that the rocket dynamics problem would be couched. This in turn implies that

$$\frac{\partial W_c}{\partial A_n} = \sum_{v=1}^{W} C_v \dot{A}_n \phi_{Cvn}^2 \tag{3.50}$$

Standard texts often assume the former of these relationships, but instead of the second they merely assume that equation (3.50) represents a small quantity. This compounds the difficulties of understanding true damping behaviour.

It is now possible to substitute these relationships back into Lagrange's equation, with the following result:

$$\ddot{A}_n \sum_{r=1}^{j} M_r \phi_{rn}^2 + \dot{A}_n \sum_{v=1}^{W} C_v \phi_{Cvn}^2 + A_n \sum_{g=1}^{S} K_g \phi_{\Delta n}^2 = \sum_{r=1}^{j} F_r \phi_{rn} \tag{3.51}$$

When this equation is compared with equation (3.8), it can be seen that there are striking similarities, and that in fact what we have done is to define the influence functions in terms of equivalent modal properties as follows:

- Equivalent or modal mass

$$\sum_{r=1}^{j} M_r \phi_{rn}^2 \tag{3.52}$$

- Equivalent or modal damping

$$\sum_{v=1}^{W} C_v \phi_{Cvn}^2 \tag{3.53}$$

- Equivalent or modal stiffness

$$\sum_{g=1}^{S} K_g \phi_{\Delta n}^2 \tag{3.54}$$

- Equivalent or modal force

$$\sum_{r=1}^{j} F_r \phi_{rn} \tag{3.55}$$

The important point of the foregoing derivation is to demonstrate that the normalization process allows the establishing of modal properties associated with each mode of vibration. Equation (3.51) is also the basis for many types of theoretical appraisal of structures. These may include the use of finite elements, matrix algebra, or numerical integration techniques. Such techniques will be passed over briefly in this book. There are plenty of alternative texts for those who are interested in pursuing such ideas.

The establishing of modal properties for a structure is of extreme importance. Theoretical techniques that do not take account of the paucity of information about some of the parameters involved in equation (3.51) do a disservice to their proponents, because they allow an illusion of precision to persist, whilst the inaccuracy involved in the estimation of individual parameters makes the result obtained somewhat debatable.

When dealing with the real behaviour of structures, the analysis using the Laplace transform contained in the former part of this chapter is of far greater significance.

Equation (3.51) is the governing equation for the dynamic response of a structure. It is now time to 'square the circle', and look back at the conventional approach to structural design. In practice, all reference to the middle term, the one that describes damping, is omitted. There is also a tacit assumption that the response occurs only at a very low frequency. (It could not be zero frequency or there would be no motion at all.) In this 'quasi-static' approach, forces and stiffnesses are equated floor by floor.

Figure 3.4 shows how the equations can be assembled for the 'quasi-static' approach. As can be seen, this leads to a truncated formulation of equation (3.51), in which the equations are:

$$\begin{array}{llllll} m_1\ddot{y}_1 & + k_1 y_1 - & k_2(y_2 - y_1) & & - & F_1(t) = 0 \\ m_2\ddot{y}_2 & + & k_2(y_2 - y_1) - & k_3(y_3 - y_2) & - & F_2(t) = 0 \\ m_3\ddot{y}_3 & & + & k_3(y_3 - y_2) & - & F_3(t) = 0 \end{array} \tag{3.56}$$

which can be written as

$$[M]\{y\} + [K]\{y\} = \{F\} \tag{3.57}$$

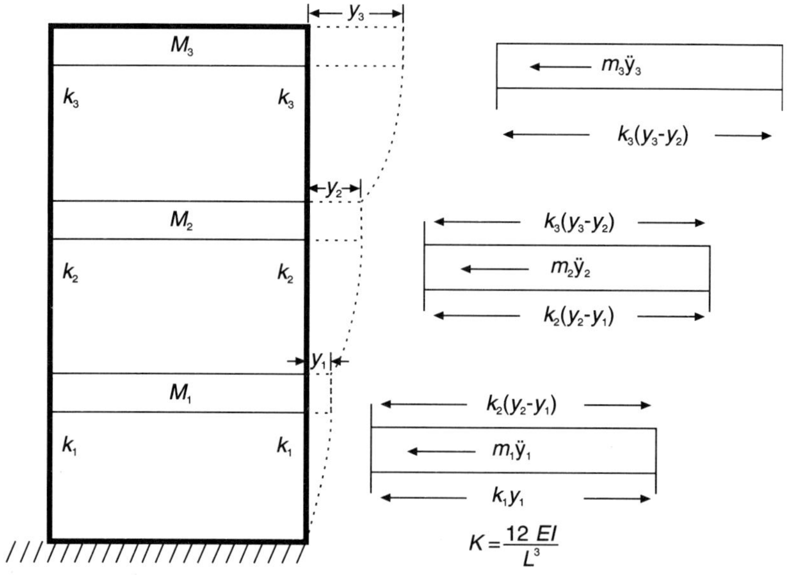

Fig. 3.4 The quasi-static approach, leading to matrices [*M*] and [*K*].

where

$$[M] = \begin{bmatrix} m_1 & 0 & 0 \\ 0 & m_2 & 0 \\ 0 & 0 & m_3 \end{bmatrix} \tag{3.58}$$

and

$$[K] = \begin{bmatrix} (k_1 + k_2) & -k_2 & 0 \\ -k_2 & (k_2 + k_3) & -k_3 \\ 0 & -k_3 & k_3 \end{bmatrix} \tag{3.59}$$

In this approach, matrices for $[M]$ and for $[K]$, the mass and stiffness respectively, are assembled. The equation is completed by assembling a force vector. This force vector is a function of time and height. As an alternative, this can be thought of as an influence function and a modal force.

In practice, it would be possible to assemble a matrix for damping as well, and if this were the case, the two approaches would be identical.

CHAPTER 4

Mode shapes, modal characteristics and influence lines

Each resonance condition of a structure has associated with it a deflected shape known as a 'mode shape'. When a structure is vibrating, it chooses a deflected form that will represent a minimum-energy solution to the dissipation of the energy to which it is subjected. Because they are minimum-energy solutions, mode shapes are consistent, and similar mode shapes are found in similarly shaped structures.

The mode shapes associated with particular modes of vibration form part of the systematic solution to the dynamic equations of motion. As such, they are part of the systems equations, which can be reversed so as to show the relationship between forcing and structural response. When used in this form, the mode shapes are an indication of the influence that a force will have on the response of a structure. Mode shapes and influence lines are therefore different aspects of the same systems approach; they merely refer to different processes. A mode shape represents the response of a structure to a forcing function, whilst an influence line represents the effect that force will have on the response.

This equivalence occurs because the greatest effect of a force on the response in one mode of vibration will occur if that force is applied at the position at which the deflection in the mode will be the largest.

It is necessary to look at the mode shapes associated with each type of structure for different types of force regime, since they dictate how the force will act on the structure. From this analysis, it can be seen that modal mass, modal damping and modal force are simply defined. These are as follows:

- Force

$$F_{\mathrm{t}} = \int F(x, y, z)\phi(x, y, z)\,\mathrm{d}x\,\mathrm{d}y\,\mathrm{d}z \tag{4.1}$$

- Mass

$$M_{\mathrm{t}} = \int M(x, y, z)\phi^2(x, y, z)\,\mathrm{d}x\,\mathrm{d}y\,\mathrm{d}z \tag{4.2}$$

- Damping

$$\zeta_{\mathrm{t}} = \int \zeta(x, y, z)\phi(x, y, z)\,\mathrm{d}x\,\mathrm{d}y\,\mathrm{d}z \tag{4.3}$$

Table 4.1 Factors to be considered and their importance

	Wind	*Wave*	*Traffic*	*Earthquake*	*Human*	*Machinery*
Tall buildings	DAORT	N	S	SDTO	FD	F
Chimneys	DAORT	N	S	SDO	N	F
Bridges	AORT	S	TOR	STO	F	F
Dams	AOR	W	TOR	SOR	N	F
Offshore structures	DW	WAORT	N	SDTOW	N	F
Floors	N	N	S	SO	FD	F

Key:
N Not an important consideration
D Response dominated by fundamental mode of vibration
A Aerodynamic admittance function must be assessed
O Orthogonal fundamental mode of vibration must be assessed
R Random force must be summed over the surface of the structure
T Torsional behaviour must be assessed
S Soil–structure interaction must be considered
W Wave–structure interaction must be considered
F Single-frequency forcing; check on 'tuning' effect

Table 4.1 summarizes the most important considerations for these combinations.

The implications of these equations are interesting, and they reduce to forms that are quite simple to use. In each case, it is normally assumed that simplifications can be made about the structure. The x, y, z functions in equations (4.1) to (4.3) above represent the three physical dimensions, and show that force, mass or damping must be summed in a three-dimensional sense. It is rarely necessary to perform such an elaborate analysis because the dimensions of structures are often continuous in one or more of the dimensions. For instance, it is rare that the width of a bridge would change along its length. Integrating in this direction reduces to a simple algebraic summation. The following is a list of the simplifications that can normally be made.

4.1 Tall buildings

Tall buildings and chimneys can be considered together. Since they have a height dimension that is significantly more than either width or breadth (hence the word 'tall'), they are often considered to be 'line-like'. This implies that the mass of the structure can be represented by a constant as height increases. In practice, the mass at each level reduces, as the requirement for carrying dead load decreases with height, and so another type of approximation is made. In this, the mass at each convenient height is lumped, and the normalized modal displacement associated with that height is used. The calculation then becomes:

$$m_r = \sum_{i=1}^{j} m_i \phi_{rn}^2 \tag{4.4}$$

where i is each convenient height level. For tall buildings, this is each floor level.

It is interesting to consider that the mass per unit height of a tall building is a constant. In this case, we may re-form equation (4.2) in the following form:

$$m_r = \int_0^H m(z)\phi_r^2(z)\,\mathrm{d}z \tag{4.5}$$

For the fundamental mode of tall buildings, the mode shape can normally be represented by a straight line, for which

$$\phi(z) = z/H \tag{4.6}$$

This mode shape is depicted in Fig. 4.1.

We may now evaluate the integral in equation (4.5) for the condition of a fundamental mode shape of a tall building, by applying equation (4.6). In this case we have

$$m_r = \left[\frac{m(z)z^3}{3H^2}\right]_0^H \tag{4.7}$$

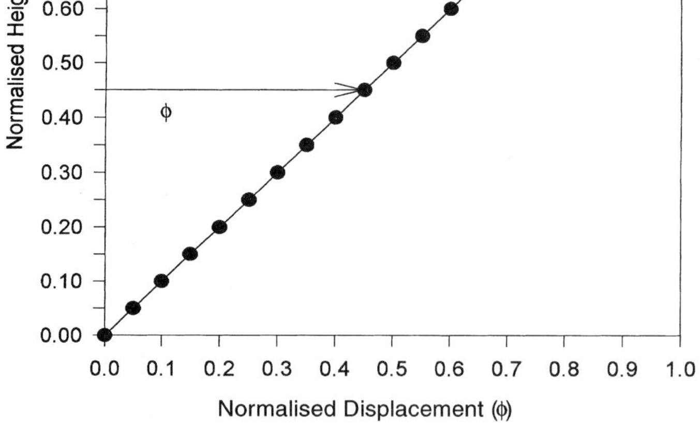

Fig. 4.1 Normalized fundamental mode shape for a tall building.

Thus

$$m_r = m(z)H^3/3H^2 \tag{4.8}$$

and noting that $m(z)H$ is the total mass of the building (M_t), we can express equation (4.8) as

$$m_r = m_t/3 \tag{4.9}$$

In other words, the modal mass is one-third of the total building mass when the deflected mode shape is represented by a straight line, as in Fig. 4.1.

This expression is of such great use that it is often applied as an approximation even when the mass per unit height deviates from a constant value. The alternative is to lump masses in the way shown in Table 4.2. Even in this case, it can be seen that the approximation of a modal mass being one-third of the total mass would lead only to an error that is less than 10%.

In some other circumstances, the mode shape may be considerably different from a straight line. In such circumstances, other expressions for ϕ should be used.

Generalized mode shapes for tall buildings are presented in Figs 4.2 to 4.6. These deflected forms represent the peak dynamic deflection against height most often encountered in tall buildings. Because the deflection is a dynamic one, it is necessary to envisage the various points of the structure oscillating about the neutral line. This they do at the natural frequency associated with that particular mode of vibration.

In the case of tall buildings, the main exception is a mode shape described as a shear mode. Figure 4.7 shows a representation of this deflection.

A shear mode can be represented mathematically by the following expression:

Table 4.2 Lumping of masses

Height (m)	*Mass* ($\times 10^6$ *kg*)	ϕ	ϕ^2	$M\phi^2$
100	0.5	1.0	1.00	0.500
90	0.7	0.9	0.81	0.567
80	1.0	0.8	0.64	0.640
70	1.0	0.7	0.49	0.490
60	1.0	0.6	0.36	0.360
50	1.0	0.5	0.25	0.250
40	1.0	0.4	0.16	0.160
30	1.0	0.3	0.09	0.090
20	1.0	0.2	0.04	0.040
10	2.0	0.1	0.01	0.010
0	3.0	0.0	0.00	0.000
	$M_t = 10.2$ (above ground)			$M_m = 3.107$

$$\phi(z) = \sin(\pi z/2H) \tag{4.10}$$

If this expression is now combined with equation (4.5), then the following expression is derived:

$$m_r = \int_0^H m(z)\sin^2(\pi z/H)\,dz \tag{4.11}$$

Then,

$$m_r = \int_0^H m(z)\tfrac{1}{2}[1 - \cos(\pi z/H)]\,dz \tag{4.12}$$

$$m_r = m(z)\left[\frac{z}{2} - \frac{H}{2\pi}\sin\left(\frac{\pi z}{H}\right)\right]_0^H \tag{4.13}$$

$$m_r = m(z)H/2 = m_t/2 \tag{4.14}$$

In this case (a shear mode deflection) it can be seen that the modal mass is half of the total mass of the building.

The foregoing has been based on simple mathematical expressions to depict the deflected form for what is essentially a cantilever. However, for the consideration of more complex mode shapes, such as second translational ones of a tall building, or modes for structures such as dams, floors and bridges, it is necessary to adopt a somewhat more rigorous approach to

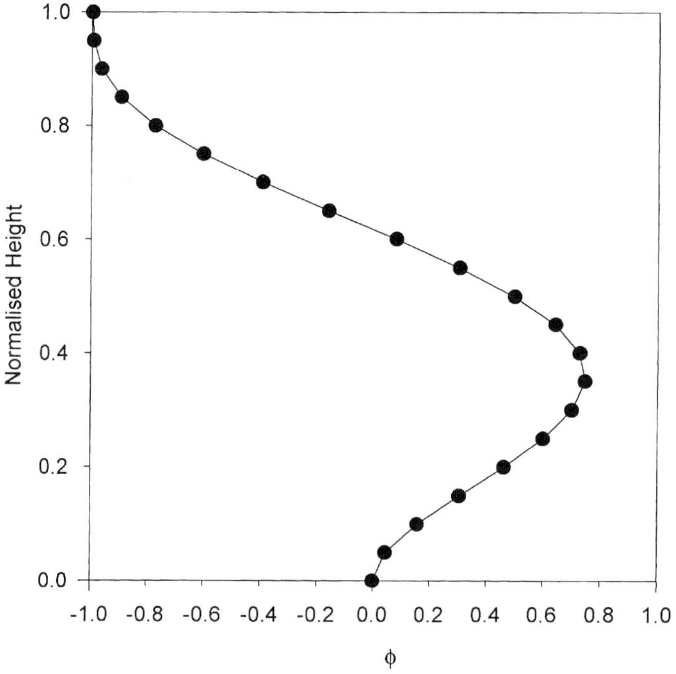

Fig. 4.2 Second translational mode for a tall building.

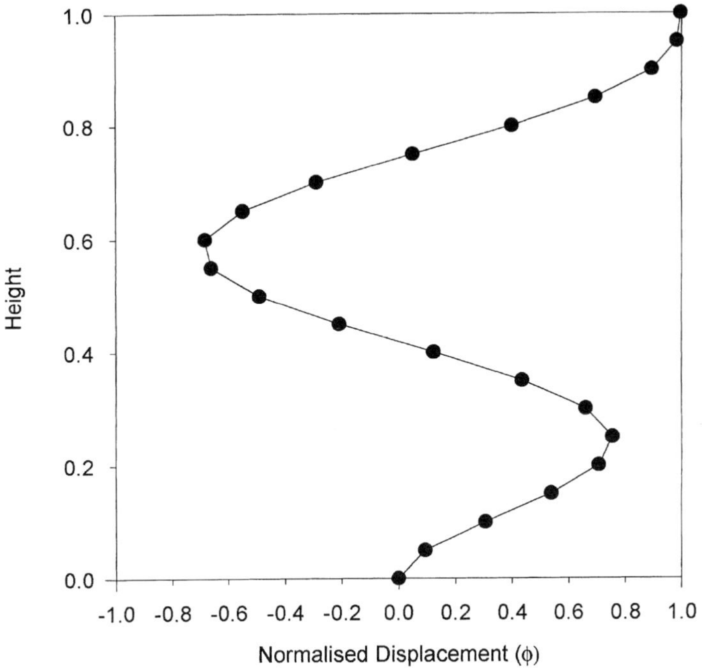

Fig. 4.3 Third translational mode for a tall building.

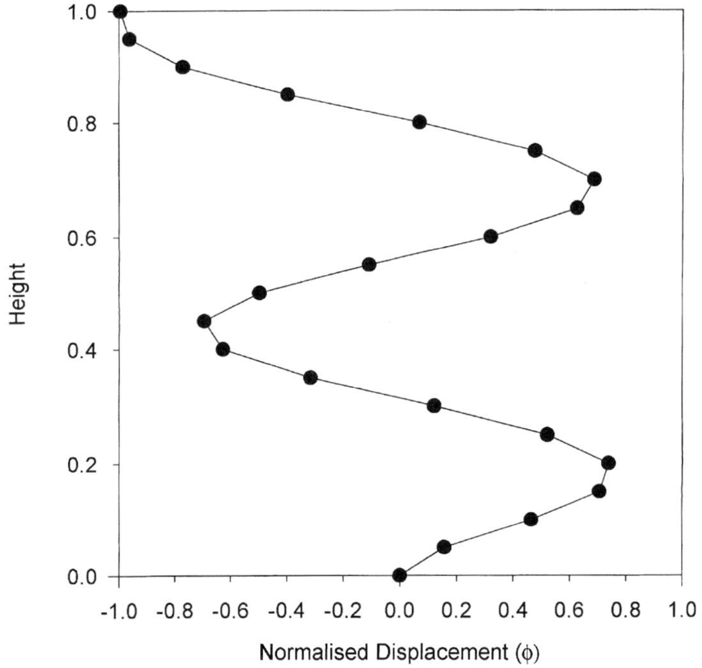

Fig. 4.4 Fourth translational mode for a tall building.

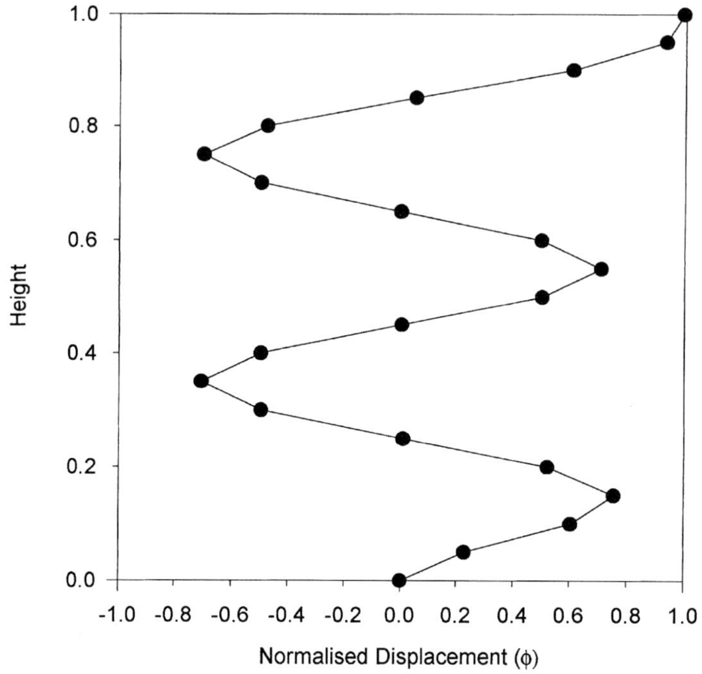

Fig. 4.5 Fifth translational mode for a tall building.

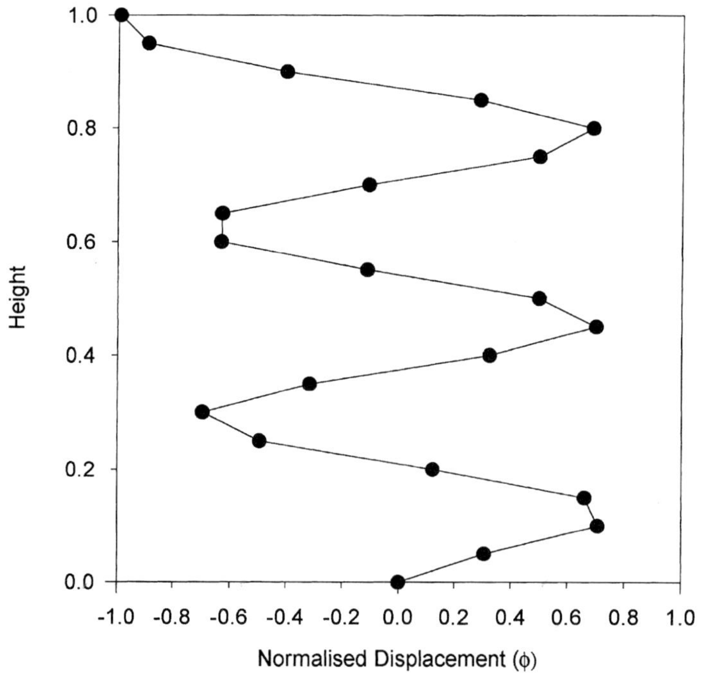

Fig. 4.6 Sixth translational mode for a tall building.

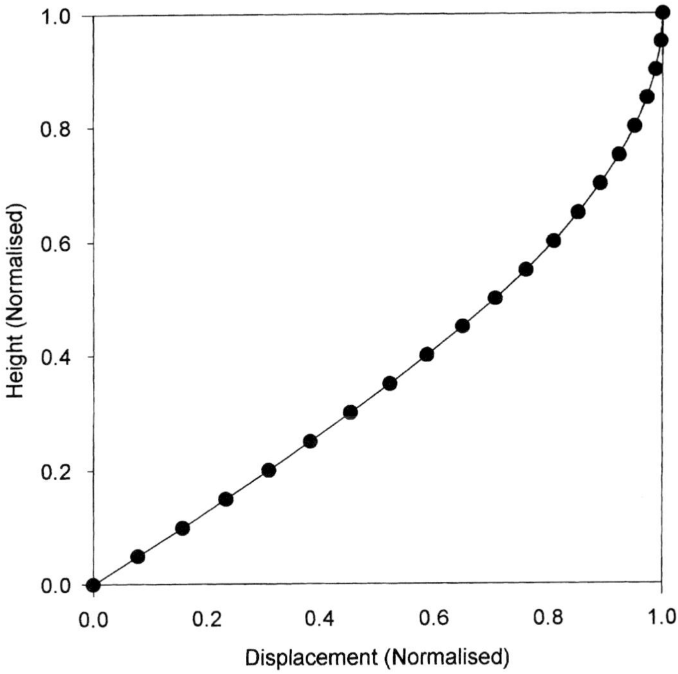

Fig. 4.7 Shear mode for a tall building.

the problem. This is because the boundary conditions now have an important role to play in the deflections of the overall structure. In this case, we may consider bending of the line-like structure, and this is governed by the following equation:

$$M = -EI\frac{\partial^2 y}{\partial x^2} \tag{4.15}$$

where EI is the rigidity of the structure. We also note that

$$F(t, z) - M''z = \frac{\partial^2 M}{\partial x^2} \tag{4.16}$$

where $F(t, z)$ is the time-varying force acting on the structure. This relationship holds if we neglect shear and rotation of elements of the structure.

This leads directly to the relationship:

$$\frac{\mathrm{d}^4}{\mathrm{d}x^4}\phi(z) - \frac{M\omega_\mathrm{n}^2}{EI}\phi(z) = 0 \tag{4.17}$$

from which, for a cantilever, a solution for the mode deflected form is

$$\phi_n(z) = K_n[\sinh(a_n z) - \sin(a_n z)] + [\cosh(a_n z) - \cos(a_n z)] \tag{4.18}$$

where $a_n = (n - 1/2)\pi$ and $n = 1$ for the first mode, 2 for the second and so on.

Additionally, for a cantilever type structure:

$$K_n = -\frac{\cos(a_n H) + \cosh(a_n H)}{\sin(a_n H) + \sinh(a_n H)} \qquad (4.19)$$

The use of these equations is quite general. In practice, equation (4.17) is the same for all types of structure, whilst equation (4.18) is modified by boundary conditions depending on the type of structure being considered.

In practice, there are two fundamental translational modes of vibration for tall buildings. It is a minimum-energy requirement that these modes are orthogonal. Orthogonal signifies that the modes are at 90° to each other. This is a mathematical requirement, not a geometrical one, although for tall buildings the effect is the same, and the fundamental orthogonal mode is set at 90° to the fundamental one.

The presence of a torsional mode very close to the two fundamental translational modes means that geometric coupling is easily accomplished by the structure, and in such cases the design for dynamic forces may be confounded by energy from all three of the fundamental modes suddenly appearing in only one of them. The designer's ability to separate such modes is easily enhanced by a decision to situate stiffness elements well away from the torsional centre of the building. In this situation, translational stiffness is maintained, whilst torsional stiffness is increased. All this is achieved with no increase in mass or translational stiffness of the structure.

The fundamental natural frequency of a tall building can be approximated by a number of different formulae based on measurements. An example of such a formula is

$$f_1 = 46/H$$

where H is the height of the building in metres.

Such formulae give a good estimate for the first natural frequency only if the building is of a 'classical' form, in other words if it resembles the buildings used in the global survey of frequencies. However, the variation can be quite large for some individual buildings. Nevertheless, the original paper that introduced this equation also showed that it was more precise than sophisticated computer-based methods such as finite-element programs. The reason for this is that such methods cannot take into account all of the non-structural elements that actually contribute, but are assumed not to in such a theoretical approach, whilst the measurements made actually sum the effects caused by these non-structural elements into the overall behaviour. The point to bear in mind is that both sophisticated and heuristic methods of estimating the frequencies of modes of vibration have some quite large confidence intervals associated with them. In the absence of measurement, a confidence interval of ±50% should be assumed.

4.2 Chimneys

The case for chimneys is very similar to that of tall buildings, and simplifications once again lead to a close approximation to mode shapes measured from real structures. However, because of the lesser amount of material in the structure, the mode of vibration is more representative of a cantilever mode. This form is observed in practice, as shown in Fig. 4.8, and can be represented mathematically as

$$\phi(z) = 1 - \cos(\pi z/2H) \tag{4.20}$$

If this mode shape expression is incorporated into equation (4.5), then we have the following:

$$m_r = \int_0^H m(z)\left[1 - 2\cos\left(\frac{\pi z}{2H}\right) + \frac{1}{2}\left(1 + 2\cos\left(\frac{\pi z}{H}\right)\right)\right] \mathrm{d}z \tag{4.21}$$

$$m_r = m(z)\left[z - \frac{4H}{\pi}\sin\left(\frac{\pi z}{2H}\right) + \frac{1}{2}\left(z + \frac{H}{\pi}\sin\left(\frac{\pi z}{H}\right)\right)\right]_0^H \tag{4.22}$$

or

$$m_r = m_z(3H/2 - 4H/\pi) \tag{4.23}$$

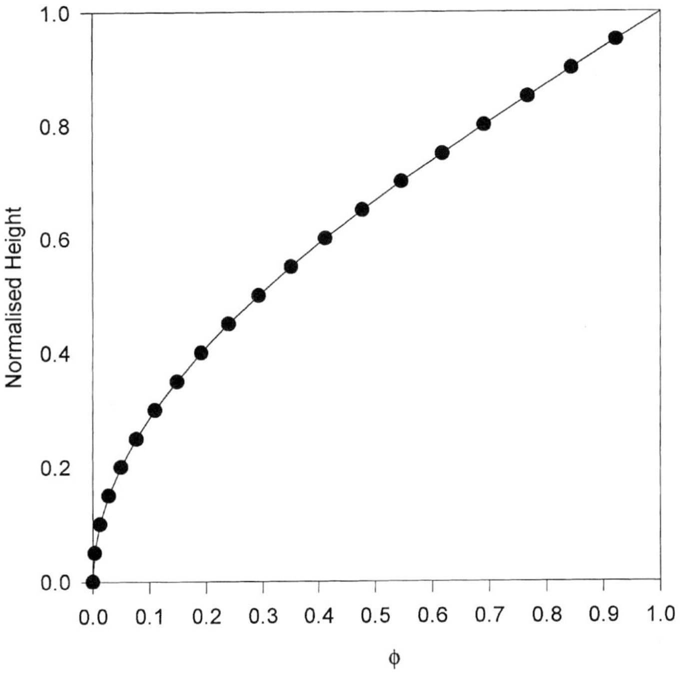

Fig. 4.8 Cantilever mode typical of modes for a chimney.

Therefore,

$$m_r = 0.227 m_t \tag{4.24}$$

or approximately

$$m_r = m_t/4 \tag{4.25}$$

Figure 4.8 is a representation of the cantilever type of mode that is typically found in chimneys.

Chimneys also exhibit orthogonal mode translational response. However, in this case, torsional mode activity is rarely of concern, since all of the stiffness is normally concentrated around the periphery. In the event that this is not the case because the wall of the chimney is not particularly thick, then shell-type modes of vibration of the chimney wall also become a consideration. Figure 4.9 gives examples of such modes of vibration. In this circumstance, complex modes involving a coupling of shell modes and bending modes complete a rich tapestry of complex modes of vibration.

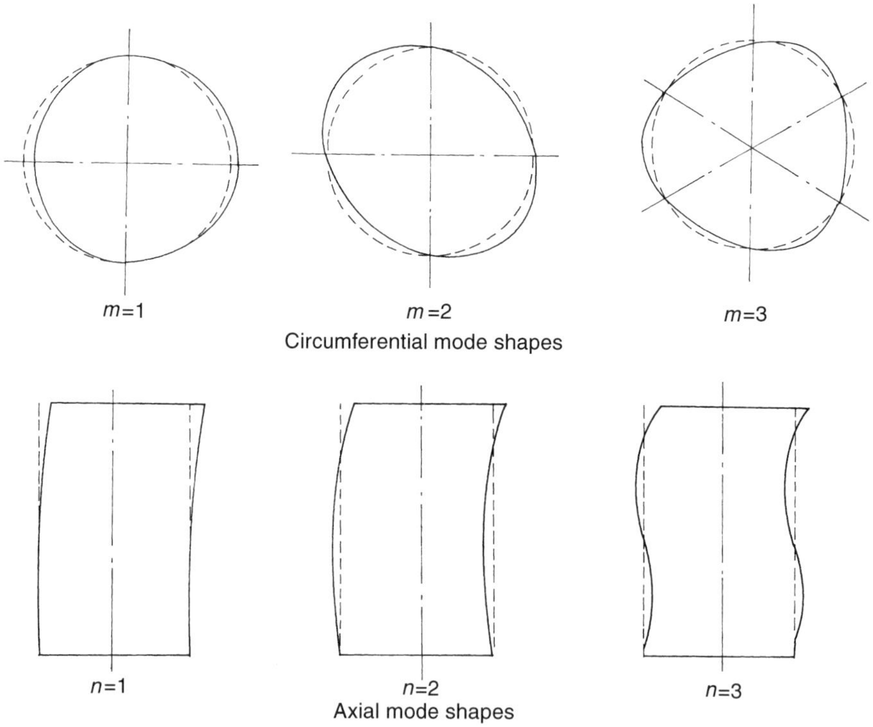

Fig. 4.9 Some modes of vibration of a cantilever cylindrical shell.

4.3 Bridges

In the case of bridges, the structure is, once again, assumed to be line-like, but the boundary conditions are different from those for tall buildings and chimneys. In this case, the fact that the structure is essentially horizontal rather than vertical has little bearing on our considerations. However, the fact that both ends of the structure are fixed is of paramount importance. (In practice, we make the assumption that the ends of a bridge are fixed rather than hinged. Neither assumption is completely correlated with observation.)

In this case we find that equation (4.18) remains the solution, but equation (4.19) is modified to become

$$K_n = \frac{\cos(a_n H) - \cosh(a_n H)}{-\sin(a_n H) + \sinh(a_n H)} \tag{4.26}$$

If we consider motion only in one direction (e.g. vertical motion), then we find a set of mode shapes of the type shown in Figs 4.10 to 4.14. These mode shapes are obtained by setting $n = 1$, 2 and 3 in equations (4.18) and (4.26), and such modes are found in practice from measurements made on real structures.

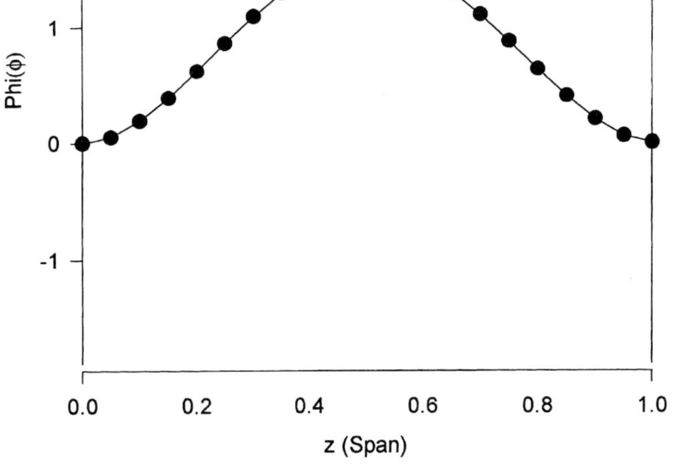

Fig. 4.10 First symmetric mode for a bridge.

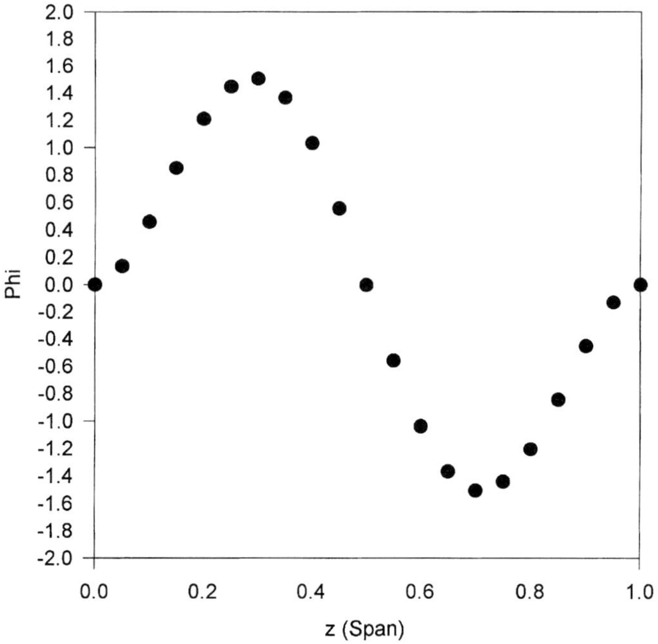

Fig. 4.11 First antisymmetric mode for a bridge.

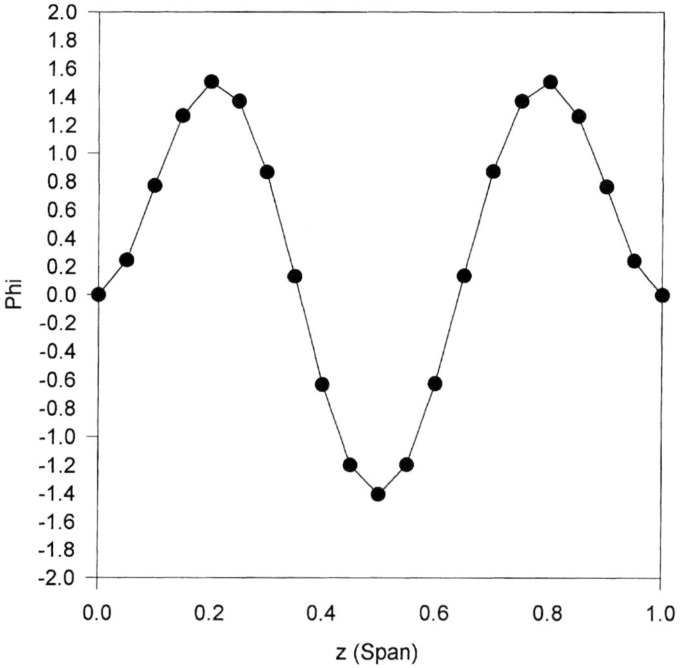

Fig. 4.12 Second symmetric mode for a bridge.

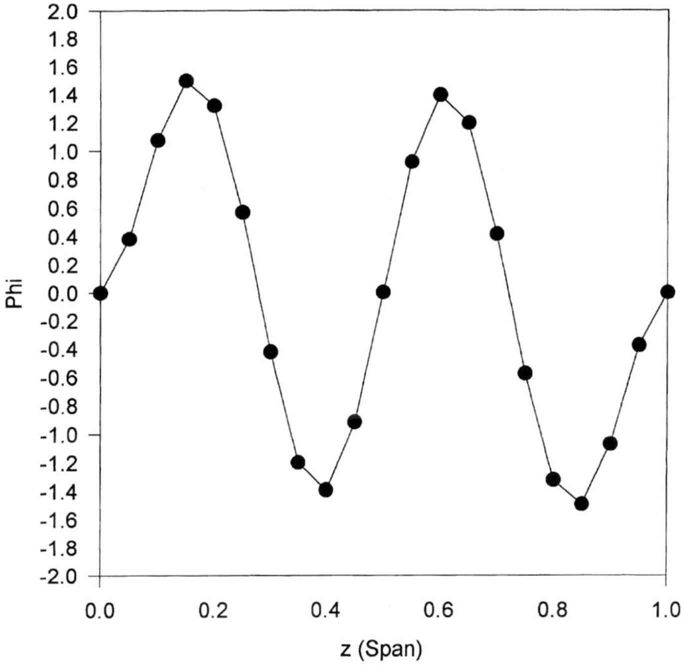

Fig. 4.13 Second antisymmetric mode for a bridge.

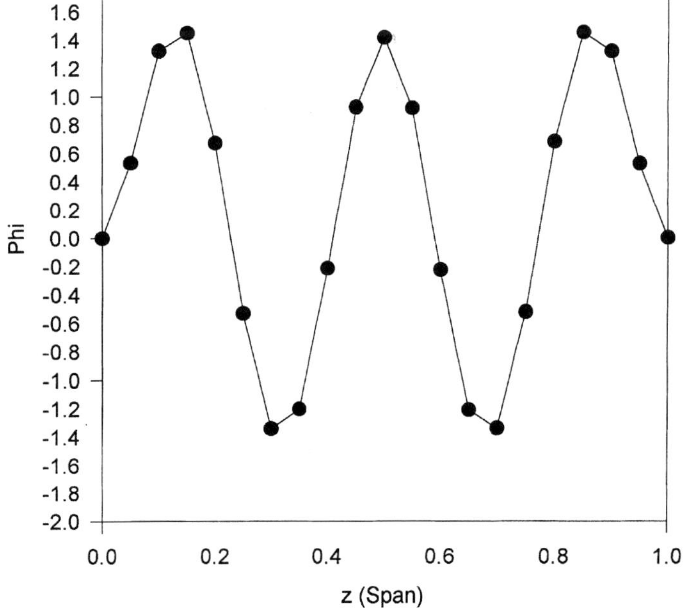

Fig. 4.14 Third symmetric mode for a bridge.

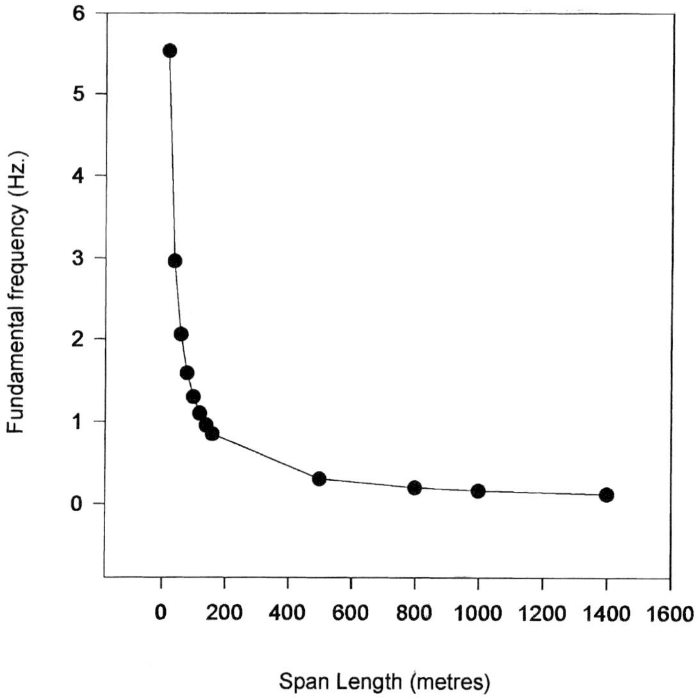

Fig. 4.15 Fundamental natural frequencies of bridges.

However, in the case of bridges, the modes may be of horizontal or vertical directionality. The set of modes shown in Figs 4.10 to 4.14 are repeated in vertical and horizontal sets. Once again, torsion of the deck is possible, and the shapes shown in Figs 4.10 to 4.14 may also be used to represent the torsional mode set. It is merely necessary to consider that ϕ represents an angle of twist at locations along the deck.

Once again, coupling between the three sets of modes is facilitated by torsional modes being close in frequency to the translational ones, and in such a circumstance energy in two of the modes can suddenly appear and add to the response in the third mode. In the case of bridge design, such considerations are of extreme importance, and the mechanical coupling can be extensively researched with a view to elimination, by the use of models, prior to construction.

The natural frequency associated with the fundamental mode of vibration of a bridge is a function of the span length of the bridge. The relationship is shown in Fig. 4.15.

4.4 Dams

There are four major types of dam:

- Concrete gravity
- Concrete arch
- Rockfill
- Earthfill

The dynamic behaviour of each type is essentially similar, and the deflected mode shapes are very similar. The major deflected forms of the modes of vibration of dams as a function of distance along the crest are similar to the mode shapes shown for bridges in Figs 4.10 to 4.14, because the boundary condition is assumed to be identical.

The major difference is that for dams they extend into three dimensions. The mode shape is, in practice, supplemented by a displacement that is a function of the height up the dam. The vertical mode shape is similar to the modes shown in Fig. 4.3 for tall buildings or for chimneys. The resulting modes of vibration are a combination of the spanwise modes (similar to bridges) and the vertical modes (similar to tall buildings). Each mode of vibration is a combination of such modal pairs. Mode shapes of dams are a strong function of the individual structure, and so measurements in detail are presented from tests on full-scale dams in Chapter 6.

The description in the previous paragraph relates to the simple form of mode shape for dams. These occur when the dam behaves as a monolith. However, it is quite often the case that a dam will behave less as a monolith when the water level in the reservoir decreases. This is because the mass of water tends to force individually constructed elements to act together. As the water level decreases, then the individual elements tend to act independently.

By this stage, it will probably come as no surprise to learn that these mode shapes are repeated in two directions. These are identified as upstream–downstream and across-valley. The basic deflected shapes are repeated in these two directions. What may be of some surprise is that a third set of modes exist in a vertical orientation. All three sets may couple together once again, and the result is a complex amalgam of three-dimensional modes of vibration.

Perhaps the most apposite visualization of such complex modes of vibration is to consider that the mass in a dam behaves in much the same way as water in a fish tank behaves, with standing waves being set up and representing resonance conditions. The existence of 'sloshing' modes of vibration (involving sloshing of the concrete) was first noticed in forced vibration tests on the Lower Glendevon Dam in Scotland in 1981.

It is also worth observing that the interface between a dam and the surrounding valley is not always clear-cut, and in several cases in which the author has been shaking a dam, the dam itself has then induced motion in the surrounding countryside. Some examples from measurements are presented later in Fig. 6.51.

4.5 Floors

The geometry of floors is relatively uncomplicated. However, boundary conditions once again are a little problematic. In calculating the mode shapes, it is assumed that the edges are fixed. It would also be possible to assume that all boundaries of the floor are simply supported. In practice, the real boundary conditions are somewhere between the two extremes.

Figures 4.16 to 4.22 show a series of modes of vibration for a typical floor. The natural frequency of a floor can be calculated as follows:

$$f_n = \frac{\pi}{2}\left(\frac{Eh^3}{12(1-\nu^2)m}\right)^{1/2}\left(\frac{j^2}{a^2}+\frac{i^2}{b^2}\right) \quad (4.27)$$

By selecting integer values of i and j, the modes of vibration are predicted.

If it is assumed that it is not possible for people to jump up and down at more than 5 Hz, then a pragmatic design criterion can be established. The most stringent design requirement occurs when the loading coincides with a

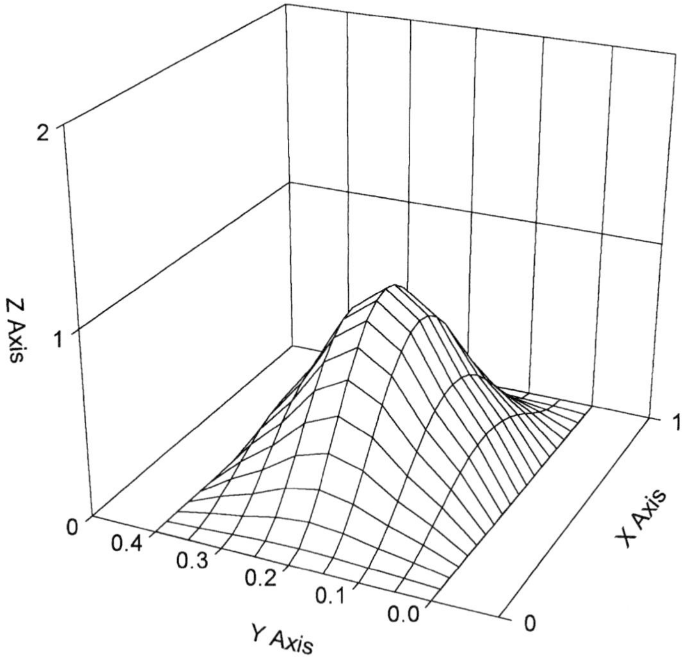

Fig. 4.16 Fundamental mode of a floor.

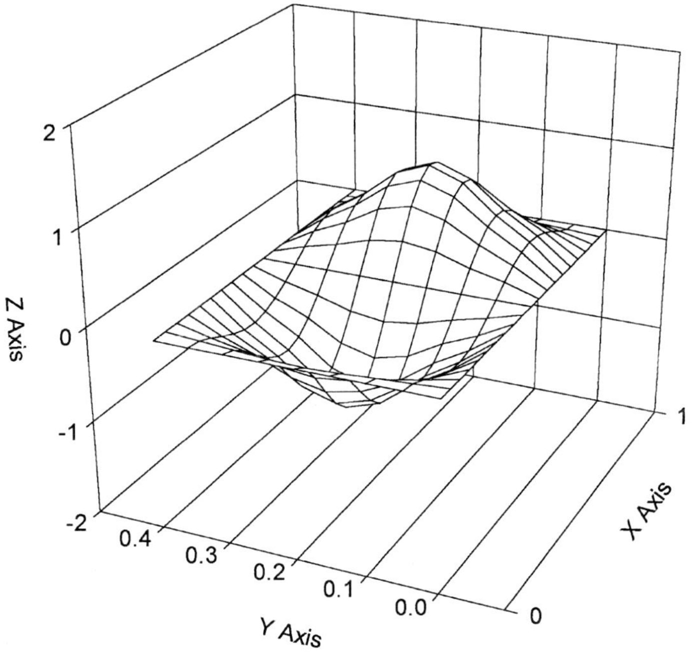

Fig. 4.17 First antisymmetric mode of a floor.

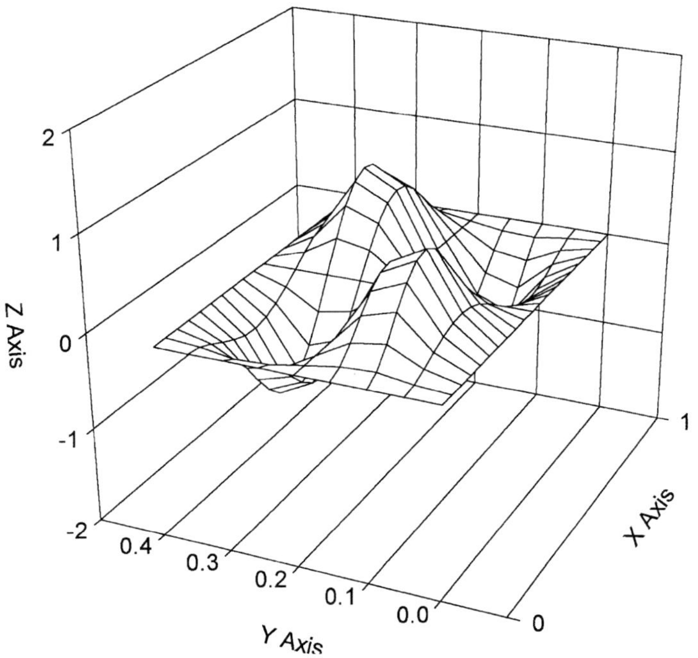

Fig. 4.18 The $i = 2$, $j = 2$ mode of a floor.

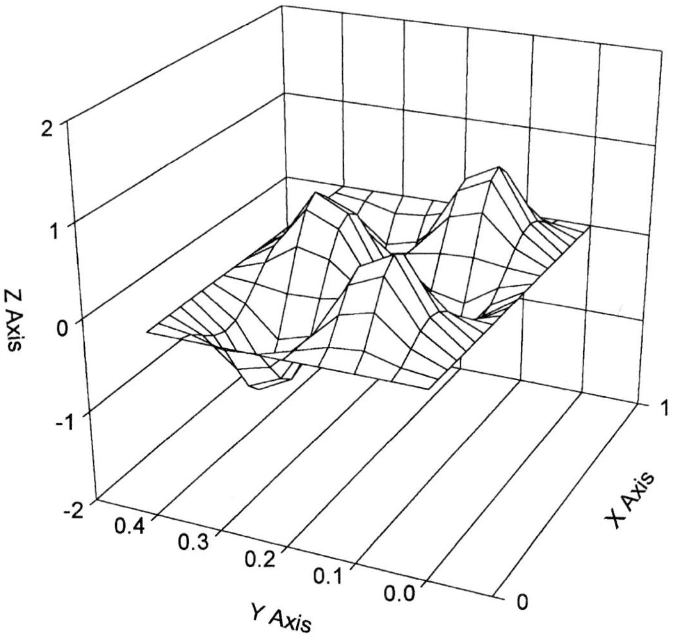

Fig. 4.19 The $i = 2$, $j = 3$ mode of a floor.

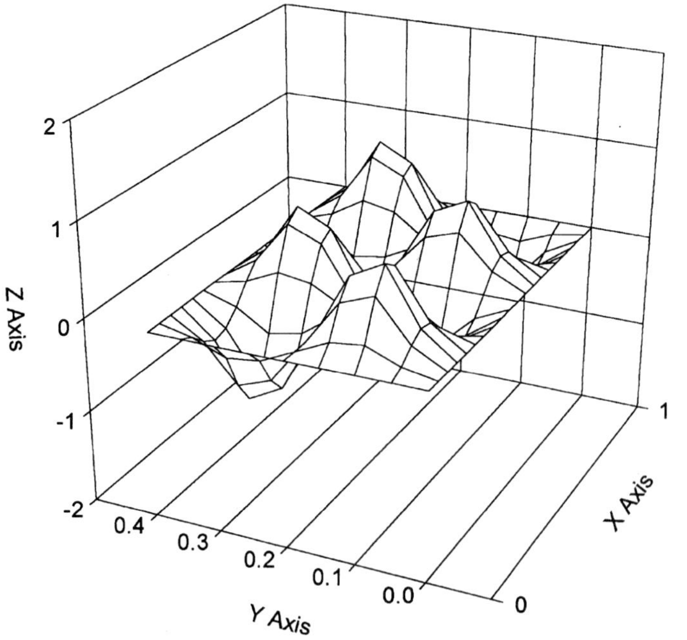

Fig. 4.20 The $i = 2$, $j = 4$ mode of a floor.

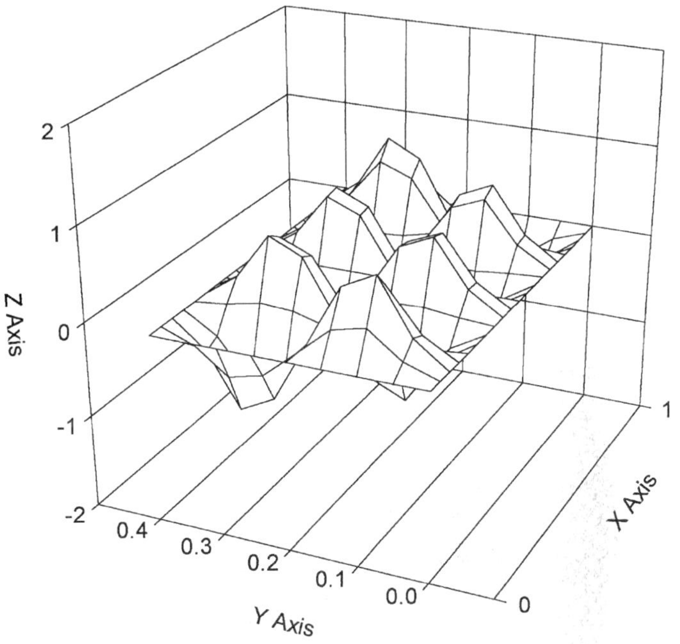

Fig. 4.21 The $i = 2$, $j = 6$ mode of a floor.

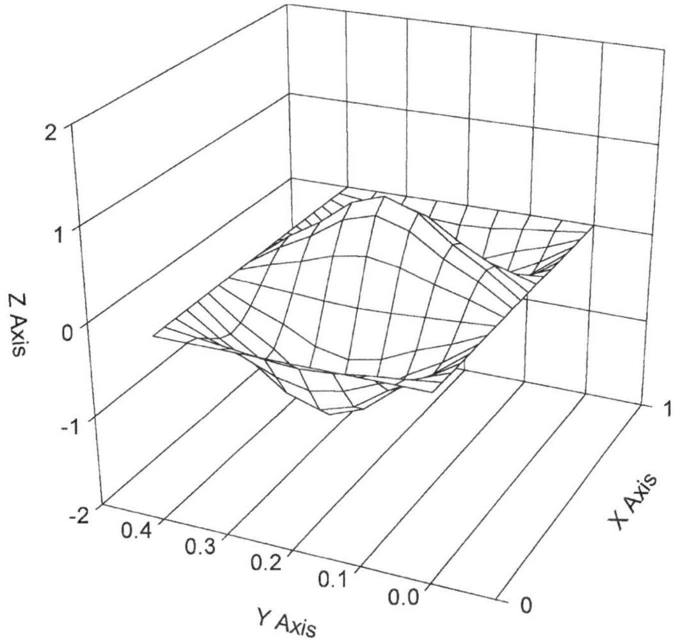

Fig. 4.22 Second symmetric mode of a floor.

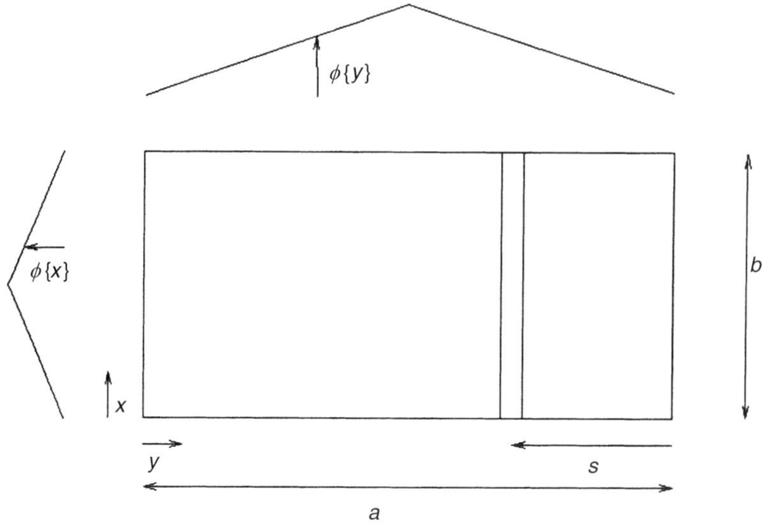

In the orientation shown
ϕ{x}=0.5
ϕ{y}=2*s*/*a*
F = 3000 x number of people

Fig. 4.23 Calculation of the modal mass of a floor.

resonance of the floor. As an alternative, the loading can be at one-half or one-quarter of the resonance frequency of the floor. Using the Duhamel integral for a parametric study, it is found that there is an effective modal influence function given by Table 4.3.

The response of the floor is then calculated from:

$$x_1(\phi) = \frac{F_1(\phi)I(\phi)}{8\pi^2 f_1^2 m_1 \zeta_1} \tag{4.28}$$

The modal force $F_1(\phi)$ is estimated from $F_1\phi(x)\phi(y)$; F_1 may be estimated as being (3000× the number of people) newtons of force; f is calculated from equation (4.27); and m is the modal mass, which may be estimated by

Table 4.3 Effective modal influence function I(ϕ) for floors

Lowest frequency of resonance	*Below 5 Hz*	*Between 5 and 10 Hz*	*Between 10 and 20 Hz*	*Non-resonance condition*
0.1%	1.0	0.30	0.10	0.04
1.0%	1.0	0.44	0.17	0.09
4.0%	1.0	0.60	0.26	0.26

calculating the total mass of the floor and dividing by 9. The value of ζ is difficult to assess in the absence of measurement – under such a circumstance, it is suggested that a value of 0.01 is used.

The modal influence function is actually a complex parameter and is used to take account of the various other effects that occur when an audience is jumping up and down in time with music. The parameters in Table 4.3 are taken from measurements of a crowd jumping up and down at a rock concert, and as such represents an averaging of such effects as:

- The audience cannot be 100% correlated in their timing.
- The audience tends to be in contact with the floor over half of each cycle of vibration.
- The mass of the vibrating system changes from floor plus audience, to floor only, and back to floor plus audience in each complete cycle.
- The audience itself dissipates energy and so contributes to the damping effect, through such mechanisms as ankle rotations, and heel slamming.

The presence of this influence function serves to remind the designer that a purely theoretical approach is incapable of taking into account all of the small non-linearities involved in this process. However, when all of these non-linearities are summed the effect can be quite large.

CHAPTER 5 *A unifying theme*

It has been noted in Chapter 3 that all of the major engineering fields of study can and do effectively use the systems approach. The major differences are involved with a characterization of the force input to the structure. Additionally, as we have seen from a study of mode shapes and the relationships between modal and structural characteristics, the effect of a force on a structure is influenced by the position in which it acts.

In this chapter we will now consider several different fields to determine the normal methods of characterization of forces and the way in which such characterizations are normally used in the systems approach.

The starting point will be concerned with wind engineering. In this case there is a curious amalgam of professions inputting information to the process of design, and as a consequence there is considerable rivalry amongst competing methods. However, when the competing methods are considered in systems terms, it is quickly seen that they are all products of a similar philosophy.

5.1 Wind engineering

We start from the original systems concept:

$$\text{displacement} = \text{force} \times \text{system}$$

The 'system' is our structure. In this case we represent each element by a spectrum or a transfer function:

$$S_{xx}(\omega) = S_{ff}(\omega) \times H^2(\omega) \tag{5.1}$$

where

$$H(\omega) = \frac{1/K}{1 - \omega/\omega_\mathrm{n} + \mathrm{j}2\zeta\omega/\omega_\mathrm{n}} \tag{5.2}$$

We have data about the wind speeds, and these must be converted to forces:

$$F(x) = \overline{F} + f'(t) \tag{5.3}$$

This signifies that the total force of the wind is composed of a mean and a fluctuating part. The mean force is

$$\overline{F} = \tfrac{1}{2}\rho C_D A V^2 \tag{5.4}$$

and the fluctuating part is

$$f'(t) = (C_D \rho A V) V(t) \tag{5.5}$$

The description of wind forces must, then, be addressed in terms of its mean strength and of its 'gustiness'. It is this latter part that has particular significance for the dynamic behaviour of structures, since the occurrence of gusts involves pulsating energy inputs to a structure. These pulsations have a frequency content dependent on the size and speed of the eddies. The relationship between this energy input and the building is ideally suited to the use of the systems approach, since we are dealing with random processes and so spectra are particularly useful.

The attempts to describe these characteristics of the natural wind were first formalized in 1756, when John Smeaton arranged for some large plates to be mounted on the side of the Forth Bridge. He monitored the deflection of these devices and established the fact that the force produced was proportional to the square of the velocity of the wind for a plate of constant area. Indeed, the formula that he produced bears a striking resemblance to many that appear in the most modern and up-to-date codes of practice.

The highest speed ever measured was that registered at the USAF airforce base at Thule, Greenland, at which a peak gust of 92.5 m/s was monitored. It is generally reckoned that the windiest location on Earth is Commonwealth Bay and George V Coast, Antarctica. At this location, a daily mean wind speed of 48.3 m/s was monitored for the 21–22 March 1951.

Most countries maintain records appropriate to their local area. The Royal Observatory in Hong Kong, for instance, has kept records of measurements of surface wind speeds for more than 100 years. The original purpose of monitoring the wind speeds was to try to predict the forthcoming occurrence of very strong winds, which could be dangerous to shipping. This situation is quite typical of the early impetus for the monitoring of wind activity, and in turn reflects the fact that winds over the oceans are stronger than those over land areas. Early measurements were thus more likely to be instituted by sea-faring nations.

High surface winds are themselves a function of entire weather systems tracking over the surface of the planet or of locally induced phenomena.

Cyclonic weather systems

These are produced by the interaction of air masses over considerable distances, or by mechanical actions induced by large mountain ranges. There is a distinction between those which occur within the tropics and those which occur between latitudes 35° and 70° (known as extratropical cyclones). Tropical cyclones generally have more energy and produce the more severe storms; they are generated in the region 8° to 20°. Cook (1989) cites four conditions as prerequisites for the generation of a typhoon (or hurricane). As well as the location on the Earth's surface (to allow the Coriolis force to effect circular motion of the weather system), the sea surface temperature must be greater than 27°C, there must be no thermal inversion until a height of approximately 10 km, and at that height there must be an inversion. Figure 5.1 shows such a process occurring. The rising cloud mass can be clearly seen, and the outward spread as it encounters a thermal inversion is also apparent.

A tropical cyclone will then continue accumulating energy as it passes over warm water. Tropical cyclones are typically 400 to 800 km in diameter, and normally will pass a point on the Earth's surface in about two days. The highest wind speeds normally occur over a period of about 6 h. A tropical cyclone does not decay immediately it reaches land or when the water temperature drops below 27°C. Observation suggests that tropical cyclones will decay into tropical depressions when the sea surface temperature drops to 23°C. Extreme gusts have been measured in the region of 70 m/s in Hong Kong, Japan and Florida (Peterka and Shahid, 1993). Figure 5.2 shows a satellite photograph of Typhoon Sally on 8 September 1996 as it passed through the South China Sea, and Figs 5.3 and 5.4 show plots of its track until it reached land, and decayed because of the increase in friction and the reduction in the availability of energy from the sea surface.

Fig. 5.1 A storm head and a thermal inversion.

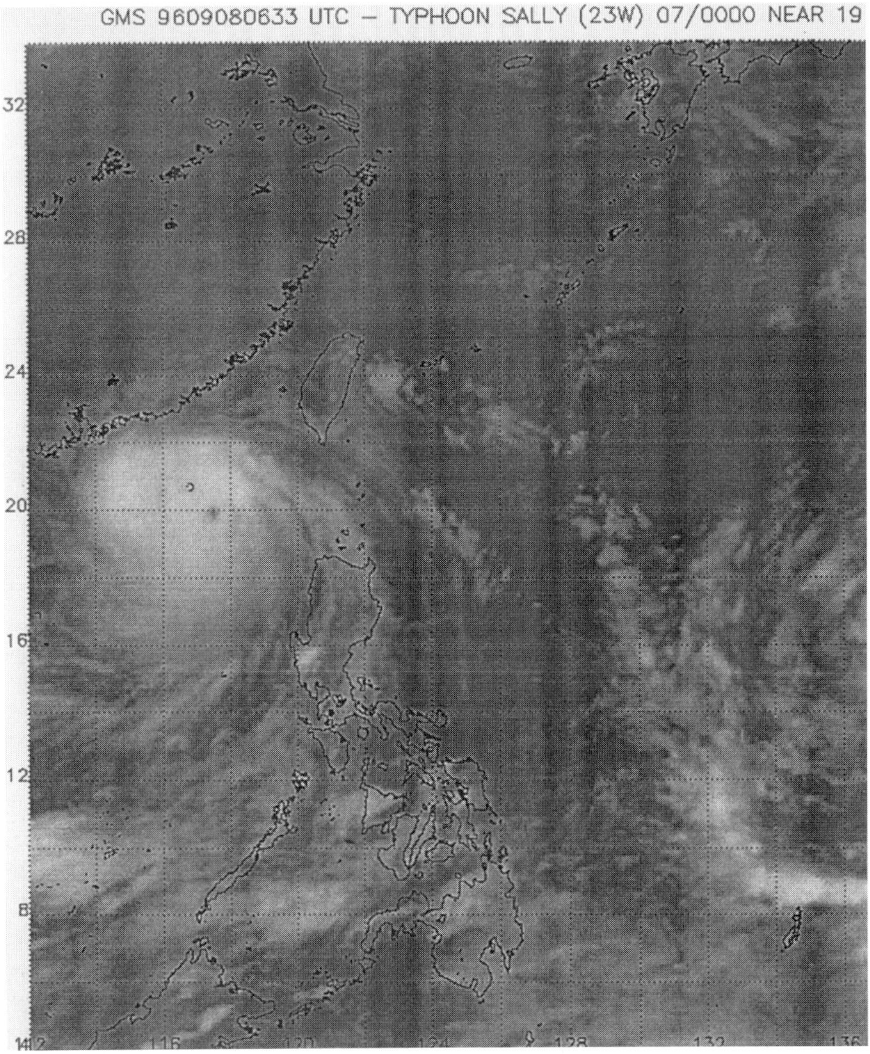

Fig. 5.2 Satellite photograph of Typhoon Sally, 1996. (Taken off the Internet. Source: Ames Research Centre, NASA.)

Tropical storms in the Atlantic, having lost their identity, can, because of their heading, take on the form of an extratropical storm.

Fronts and depressions

In more temperate regions, large masses of cold and warm air interact and produce weather systems that are typically 4000 km in diameter. In the

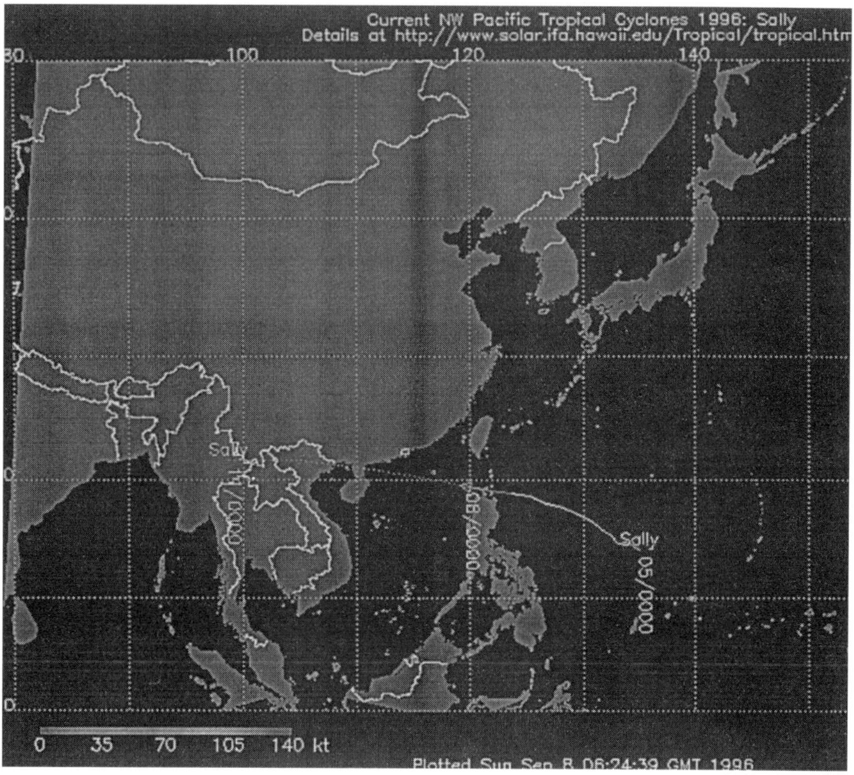

Fig. 5.3 The track of Typhoon Sally. (Taken off the Internet. Source: US Military Base, Guam. Reproduced with permission.)

Northern Hemisphere, warm air moving northwards from the subtropical regions, encountering polar fronts, can produce extremely intense storms, such as that in October 1987 over the north-west of Europe in which gusts reached 45 m/s. Figure 5.5 shows such a front in North America, in which the boundary between warm and cold air masses is delineated by a visible squall area.

Tornadoes

Typically, a tornado is 100 to 300 m in diameter, and their touch-down tracks are of the order of 15 to 20 km. Typical values of wind speed are 80 m/s, although the extreme maxima are very difficult to assess.

The most common cause of generation of tornadoes is associated with thunderstorm activity, although hurricanes can also spawn them. The mechanism of creation is through a rising column of air, which draws in air around the base to form a vortex. When this reaches a region high up where

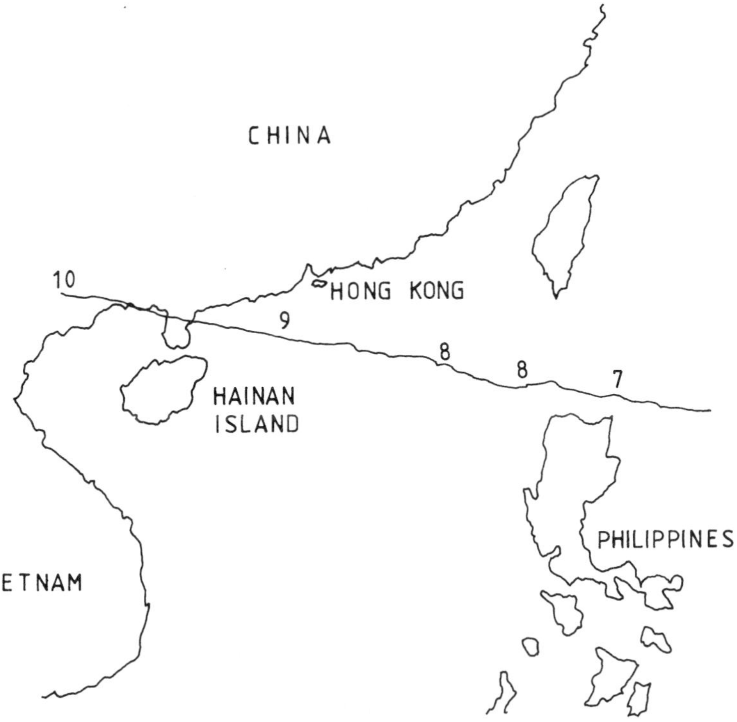

Fig. 5.4 The landfall of Typhoon Sally, September 1996.

Fig. 5.5 A squall at the boundary of warm and cold air.

jet streams exist, then the central pressure suddenly drops, producing a typical rotating vortex.

Thunderstorms

Thunderstorms inevitably involve the transfer of large quantities of energy from one form to another. The mechanism for the creation of strong winds within a thunderstorm is through the viscous drag of very large quantities of water droplets on masses of air. This can create a strong down-draught, or even a downburst. Peak winds approaching 75 m/s are possible, and in the case of a 'supercell', 'derecho', or mesocyclone, the effects can spread over an area of 1500 km^2. In a recent event in Canada (Cummine *et al.*, 1992), a 'blowdown' of trees was produced over an area of 2000 km^2.

The effects noted above are the meteorological spawning grounds for major strong winds on the surface of the Earth. There are other effects (such as lee waves and effects associated with gravity or thermally induced movements of air in mountainous regions) that can amplify strong winds or can themselves produce isolated instances of very high wind speeds.

5.1.1 Location of strong winds

The locations most affected by hurricanes and typhoons have been estimated differently by various authorities (Cook, 1989; ESDU, 1982; Jeary, 1991). The areas of influence of these strong winds differ depending on whether longer return periods are considered or whether a warming atmosphere is an influence.

5.1.2 The atmospheric boundary layer

From the point of view of the designer of structures, the mechanism for the generation of strong winds is of less importance. Interest centres on the ability to predict the description of the wind field that will affect the structure during its lifetime. Here we encounter a series of difficulties. It is possible to see that the description that is most apposite has changed throughout history. It is also possible to convince ourselves that our present description is well based in the scientific method. Nevertheless, we must describe the nature of the wind in terms of its structure at a location and its recurrence. Both of these quantities are random in nature. In addition, the descriptions used must be simplified in order that they can easily pass from one allied field to another. Perhaps the most important factor, though, is that, in keeping with the spirit of scientific inquiry, such models must be tested against full-scale measurements and must earn their status through verification.

An impetus to the measurement of the structure of strong winds was given by the quantum leap in the conceptualization of the processes operating, through work in the early 1960s (Davenport, 1961). The description of the nature of wind in terms of its structure close to the surface was the single most important concept in wind engineering during this century.

The early concepts were much associated with the description of the variation of the mean wind profile with height, the turbulence associated with strong winds, the intensity of that turbulence, the gust maximum velocity, the relationship between the mean wind speed and the peak gust, and the spectrum of wind gusts with respect to their frequency content. In addition, the size of gust eddies was of particular concern because there was an early understanding that the interaction of structures with the gusts was crucially dependent on the ratio of the structure to gust size. An appreciation of all of these factors helps in the establishment of the input force caused by the wind at a point on the Earth's surface.

5.1.3 The structure of winds and their measurement

The atmospheric boundary layer in the first 2000 m is now discussed and the parameters are considered in a little more detail, together with comments on their measurement.

Velocity and direction

Since early work in the 1960s (Davenport, 1961), there has been a realization that the gustiness of the wind must be defined if the effect on structures is to be predicted rationally. Such a description of what is a random process must of necessity be statistical in nature. Figure 5.6 gives a representation of the passing of a storm and the fundamental parameters that must be of interest in design considerations. These are the mean wind speed (V in Fig. 5.6) and the variance about the mean (σ in Fig. 5.6).

From the basic measurement of wind speed and of direction, several important parameters can be deduced, through computer analysis, as outlined below.

Intensity of turbulence

For a wind record, the mean velocity is calculated appropriate to a preselected averaging period. The standard deviation of the fluctuations of the wind velocity about this mean value is then computed. The value of the intensity of turbulence I_u is then calculated as

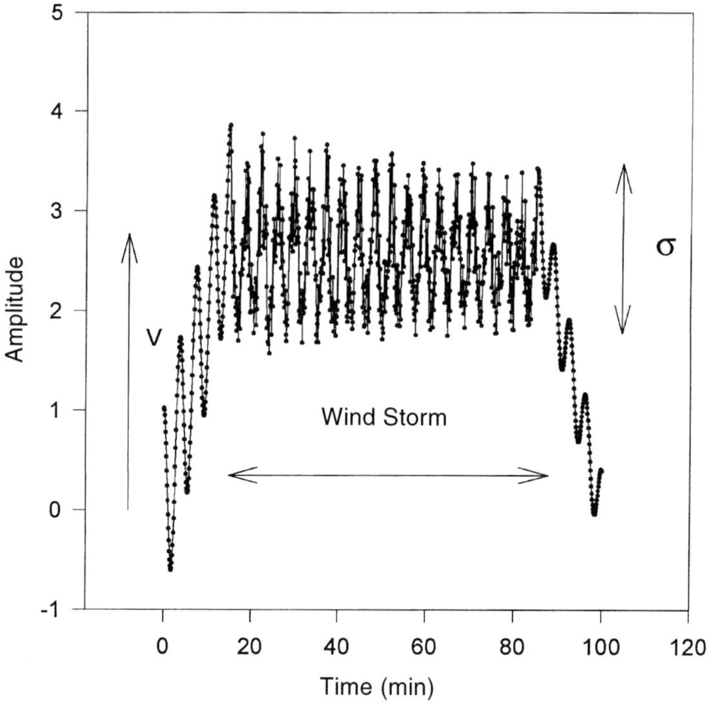

Fig. 5.6 Turbulence intensity.

$$I_u = \sigma_u / V_z \tag{5.6}$$

Gust factor

The description of the relationship between the peak gust and the mean wind speed is of considerable interest. The gust factor K_τ is simply

$$K_\tau = \hat{V}_z / V_z \tag{5.7}$$

Peak factor

The peak factor (g) is a related quantity and is expressed as

$$g = (K_\tau - 1)/I_u \tag{5.8}$$

The relationship between these quantities is dictated by the intensity of turbulence at a height h above ground, as is witnessed by the expression

$$K_\tau = 1 + gI_u \tag{5.9}$$

The provision of a record of wind data in a digital computer allows further analysis to be conducted, and a series of other measures of the statistical properties are possible, as follows.

Spectrum of horizontal gustiness

Since the energy in the wind impinges on a structure through the mechanism of an interaction of gusts of wind with the structure, then it is important to know how much energy is present as a function of frequency. A spectrum may be generated directly from windspeed data. Alternatively, the spectrum of pressures may be presented through a linear transformation, or the ordinates may be weighted by the frequency to produce a $nS(n)$ spectrum in which the power in the signal is represented as a function of frequency or of reduced frequency. The reason for presenting the data in this fashion is basically that in the $nS(n)$ form the value of the energy in the wind can be fed directly into a formulation for the calculation of the response of a structure to such a wind. Indeed, the selection of a single value from such a representation at a frequency appropriate to a resonance of a structure can be used for a direct calculation of the response in that mode of vibration. An example is the formulation:

$$\ddot{x}_{\max} = g_{\mathrm{D}}\left[\sqrt{\left(\frac{\pi}{4\zeta_1} n_1 S(n)\right)} \frac{\phi}{m_1}\right] \tag{5.10}$$

where

$$g_{\mathrm{D}} = 2\ \ln(nT) + \frac{0.577}{2\ \ln(nT)} \tag{5.11}$$

This equation implies that the response of a structure to a wind load can be thought of in terms of two basic parts. The first is essentially quasi-static and may be thought of as the response to the mean wind. The second part is the dynamic response and is the structure's response to the sub-inertial part of the spectrum at higher frequencies. The approach using equations (5.10) and (5.11) is that taken by the Engineering Sciences Data Unit in Item 76001. An example of its use is given in Appendix E.

There are two implications. As previously mentioned, the measurement of the wind speed and its representation in the frequency domain require the standard assumptions for spectral analysis to be true. The data set forming the record of fluctuating wind speeds *must* therefore form a stationary sample if the spectrum of wind fluctuations as a function of frequency is to be useful for design purposes. Secondly, the separation of the sub-inertial range of the spectrum from the inertial range must be valid. The spectral gap must exist in order to remove trends from the data sets. This spectral gap is itself assumed to exist largely as a result of investigations such as that at Brookhaven (Van der Hoven, 1957). The amount of decoupling between the microclimatological sub-inertial regime and the macroclimatological low-frequency regime is exaggerated in such plots when the ordinate is multiplied by frequency (Fig. 5.7).

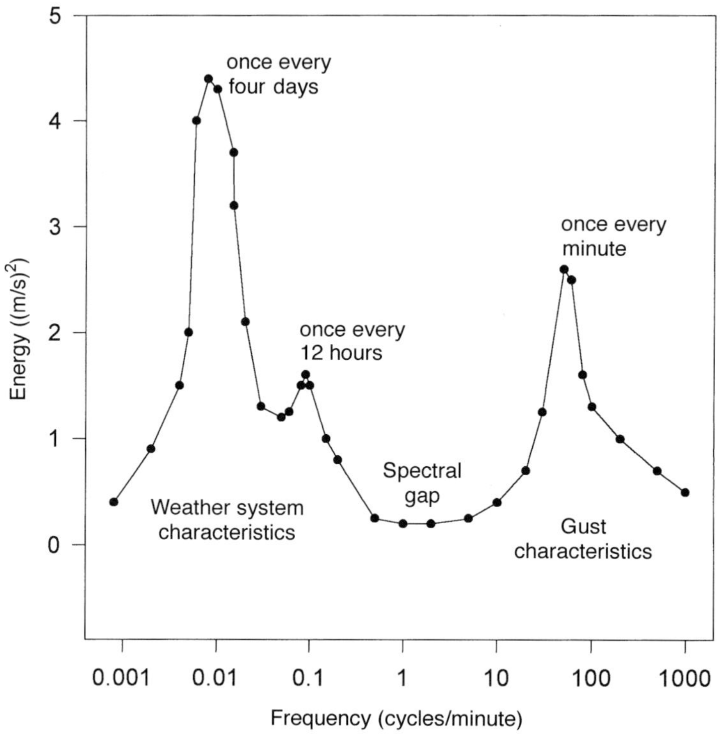

Fig. 5.7 Van der Hoven spectrum.

All of the measurements mentioned above require these two conditions before any such analyses of the data are valid. In wind-tunnel tests, the separation of the micro- and macroclimatological parts of the spectrum is induced. This in turn creates a condition in which stationarity of the remaining turbulent spectrum is much easier to achieve. The investigator in the field has no such luxury. He or she must attempt to remove the non-stationary data before beginning the analysis. It is a salutary warning that, in reviewing many of the largest studies of wind observations, the mention of tests for stationarity is one of the rarest events reported.

Vertical profiles

Measurements made at a series of vertically separated heights can provide a direct measurement of the vertical profile of the boundary layer. Measurements of velocity at different heights are compared and a profile is constructed. In this case, the time over which the velocity measurement is characterized is important. Averaging over periods such as one hour will produce (normally) well behaved vertical velocity profiles. Turbulence intensity variation with height can also be presented in this manner. If the averaging process is reduced to the region of a few seconds (normally 3 s), then a gust profile may be constructed. In the field, vertical transportation of wind gusts tends to make the gust velocity profile and the short-term turbulence intensity profile rather variable quantities. Nevertheless, they are such important parameters that several attempts have been made to measure them.

In addition to the direct measurements noted above, there are some inferences from measurements that can be made, as follows.

Integral length scales

Length scales in the x, y and z senses are an important measure of the size of gusts. There are nine integral scales, and they are of particular interest to wind-tunnel practitioners since the scale of a model should be imposed by the ratio of the prototype to full-scale integral length ratios.

The measurement of the integral length scales (${}^{x}L_{\mathrm{i}}$, ${}^{y}L_{\mathrm{i}}$, ${}^{z}L_{\mathrm{i}}$) can be achieved by integrating the autocorrelation function. The autocorrelation is the inverse Fourier transform of the spectrum of wind velocities in the x, y or z direction as appropriate.

Length scales in the x sense can be obtained from an autocorrelation function from a single point (or anemometer). This is because the autocorrelation function effectively plots the correlation of a shaft of air moving past that point with itself. A zero value of the autocorrelation function would indicate that a coherent mass of air was exhausted at that distance and would therefore effectively define the limit of the coherent gust. Unfortunately, the autocorrelation function correlates noise signals as well as the useful information, and the function often does not decay to zero. Accordingly, artificial definitions such as the time taken to decay to a value of 1/e are introduced, or corrections for the self-correlating noise may be added. Under such circumstances, it is common practice to invoke Taylor's hypothesis to convert from an integral time to an integral length.

Many difficulties have been experienced with the measurement of this parameter. This is perhaps not surprising, given the fact that it is rare to test for stationarity, but additionally the autocorrelation function averages information at all frequencies present. Considering that gust sizes are associated with different frequencies in the microclimatological peak, it is perhaps surprising that no attempts to measure integral lengths as a function of frequency appear in the literature.

Values of ${}^{y}L_{\mathrm{i}}$ and ${}^{z}L_{\mathrm{i}}$ may be obtained in a similar manner, with the exception that cross-correlations between anemometers in the vertical or the lateral sense must be performed. All other aspects of the analysis remain unchanged.

The roughness length

The roughness length z_0 can be inferred from the measurement of turbulence intensity through the relationship

$$I_u = 1/[\ln(z/z_0)] \tag{5.12}$$

The friction velocity

The friction velocity U_* can be deduced from the logarithmic relationship for a neutrally stable atmosphere:

$$U_* = \overline{V}K/[\ln(z/z_0)] \tag{5.13}$$

Surface drag

The surface drag C_D can be deduced from

$$C_D = (U_*/V)^2 \tag{5.14}$$

The parameters described above are derived from relatively straightforward relationships as noted. However, these all assume that the atmosphere is well behaved, that the data are stationary and that a stable terrain exists. Clearly there are many situations when this is not the case, and measurements from such ill-defined situations must be sought. There are many reported attempts to measure these parameters in two particularly difficult situations. These concern the flow of wind over hills, and the measurement of wind characteristics at critical landfall situations or when the boundary layer is changing its profile because of a change in roughness encountered.

Perhaps the most convenient way to describe the forces produced by the wind is by splitting the wind forces into two parts, associated with the mean wind and the higher-frequency gustiness. We then have a concept of a 'quasi-static' response and a dynamic response. The quasi-static part can be dealt with quite simply, whilst the dynamic part is most often described in terms of the spectrum of the action on a structure.

Equation (5.5) above describing the fluctuating part of the wind force can be transformed into a spectral equivalent:

$$S_{ff}(\omega) = (C_D\rho AV)^2 S_{uu}(\omega) \tag{5.15}$$

In practice, the actual value of $S_{ff}(\omega)$ drops at the high-frequency end, and so we define a parameter $\kappa^2(\omega)$, which is defined as

$$\frac{\text{Real } S_{ff}(\omega)}{\text{Ideal } S_{ff}(\omega)} \tag{5.16}$$

This parameter is called the aerodynamic admittance function, and is shown in Fig. 5.8.

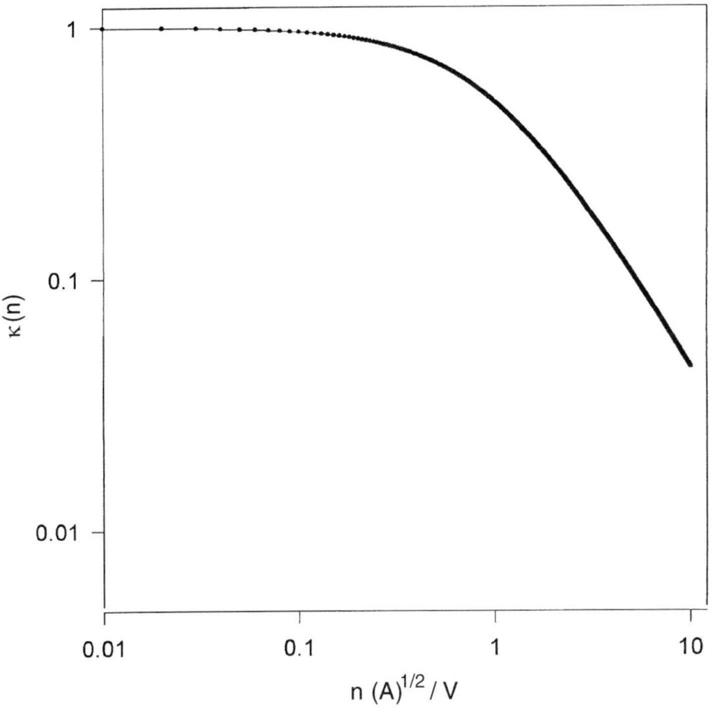

Fig. 5.8 Aerodynamic admittance function.

The equation above can now be rewritten as

$$S_{ff}(\omega) = \kappa^2(\omega)(C_D \rho A V)^2 S_{uu}(\omega) \tag{5.17}$$

Then

$$S_{xx}(\omega) = \frac{4F^2}{V^2}\kappa^2(\omega)\frac{H^2(\omega)}{K^2}S_{uu}(\omega) \tag{5.18}$$

Also

$$\sigma_x^2 = \int_0^\infty S_{xx}(\omega)\,\mathrm{d}\omega \tag{5.19}$$

and therefore

$$\sigma_x^2 = \frac{4F^2}{V^2K^2}\int_0^\infty \kappa^2(\omega)|H(\omega)|^2 S_{uu}(\omega)\,\mathrm{d}\omega \tag{5.20}$$

We note that

$$X = F/K \tag{5.21}$$

and therefore that

$$\frac{\sigma_x^2}{X^2} = \frac{4}{V^2} \int_0^\infty \kappa^2(\omega)|H(\omega)|^2 S_{uu}(\omega)\,\mathrm{d}\omega \tag{5.22}$$

or

$$\frac{\sigma_x^2}{X^2} = \frac{4\sigma_x^2}{V^2} \int_0^\infty \kappa^2(\omega)|H(\omega)|^2 \frac{S_{uu}(\omega)}{\sigma_x^2}\,\mathrm{d}\omega \tag{5.23}$$

The parts of the equation under the integral sign represent the background and the resonance parts of the response of the structure. This can be expressed as

$$\frac{\sigma_x^2}{X^2} = \frac{4\sigma_x^2}{V^2}\left(B + \frac{SE}{\zeta}\right) \tag{5.24}$$

or

$$\frac{\sigma_x}{X} = \frac{2\sigma_x}{V}\sqrt{\left(B + \frac{SE}{\zeta}\right)} \tag{5.25}$$

Graphs for B, S and E are shown in Figs 5.9, 5.10 and 5.11, respectively. For B and S the graphs are presented as a function of λ, the aspect ratio of the structure (h/L_h).

Alternatively B, S and E can be calculated from formulae (5.26), (5.28) and (5.29) below:

$$B = \frac{1}{1 + (\sqrt{36h^2 + 64b^2})/L_\mathrm{h}} \tag{5.26}$$

where h is the height of the structure, b is the breadth of the structure (normal to the wind direction) and L_h is the effective turbulence length scale given by

$$L_\mathrm{h} = (h/10)^{0.25} \tag{5.27}$$

$$S = \frac{1}{(1 + 3.5n_\mathrm{a}h/\overline{V}_\mathrm{h})(1 + 4n_\mathrm{a}b/\overline{V}_\mathrm{h})} \tag{5.28}$$

where n_a is the fundamental frequency ($= 46/h$) and $\overline{V}_\mathrm{h}$ is the mean hourly wind speed at height h

$$E = 0.47N/(2 + N^2)^{5/6} \tag{5.29}$$

where N is an effective reduced frequency given by

$$N = n_\mathrm{a}L_\mathrm{h}/\overline{V}_\mathrm{h} \tag{5.30}$$

We also note that

$$\hat{X} = \overline{X} + g_\mathrm{v}\sigma_x \tag{5.31}$$

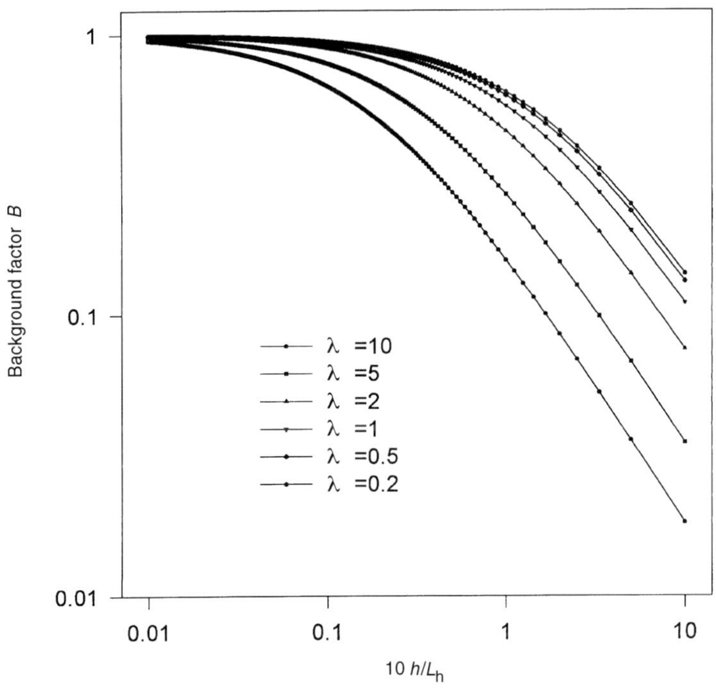

Fig. 5.9 Background factor B, as a function of $\lambda = b/h$. Curves are labelled from the top downwards. (After Vickery, 1973.)

where g_v is the peak factor. The peak factor may also be defined as

$$g_v = (\hat{V} - \overline{V})/\sigma \tag{5.32}$$

The gust factor is defined as

$$K_\tau = \hat{X}/\overline{X} \tag{5.33}$$

and

$$K_\tau = 1 + g_v I_u \tag{5.34}$$

This is the ratio of the expected peak value to the mean value and is termed the 'gust response factor', in which

$$K_\tau = \frac{\hat{X}}{\overline{X}} = 1 + 2gr\sqrt{\left(B + \frac{SE}{\zeta}\right)} \tag{5.35}$$

where

$$r = \sigma_v/\overline{V} \tag{5.36}$$

This approach was introduced in the 1960s (Davenport, 1961) and is still in dominant use today. The current Australian code AS1170 uses the

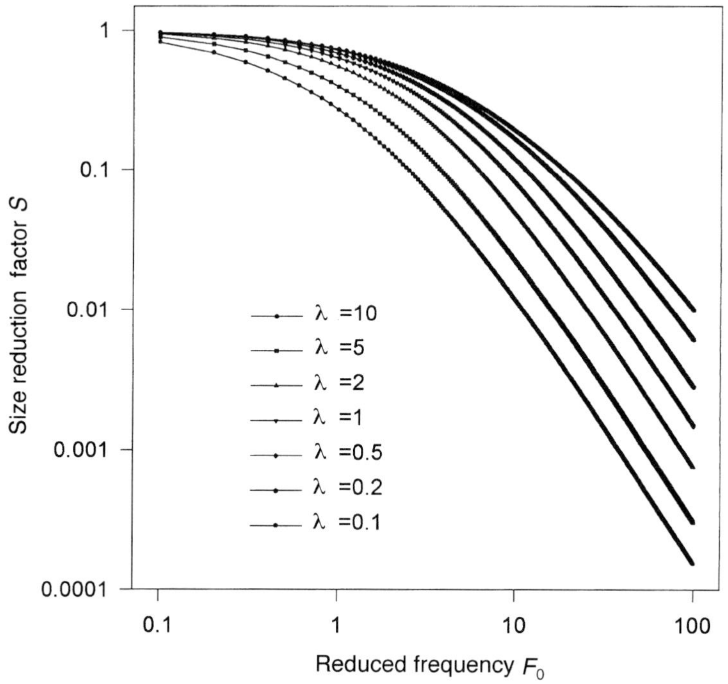

Fig. 5.10 Size reduction factor S, as a function of $\lambda = b/h$. Curves are labelled from the top downwards. (After Vickery, 1973.)

concepts outlined above. Additionally, a draft version of a new code for Hong Kong uses the same approach. Clearly, a vital component of the calculation is the establishment of a value for the damping ratio ζ. This parameter is discussed separately in Chapter 7.

Values of some windspeed variables measured for several typhoons at Waglan Island, Hong Kong, are shown in Table 5.1.

It is normally considered that the variation of wind speed with height is determined by the relationship

$$\overline{V}(z) = \overline{V}(10)(z/10)^{\alpha} \tag{5.37}$$

This relationship represents the fact that there is friction at ground level, which slows down the wind. This, in turn, induces a stress in the layers of wind, and the effect works its way further up into the atmosphere. At a height V_g, known as the gradient height, the effect of ground friction disappears completely.

It is also considered that the turbulence intensity I_u varies with height in the following form:

$$I_u = 1/[\ln(z/z_0)] \tag{5.38}$$

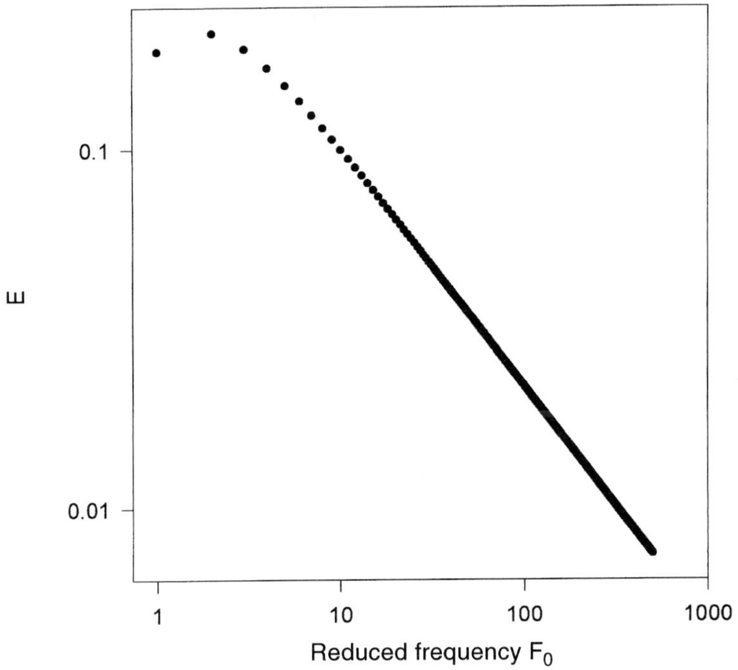

Fig. 5.11 Gust energy E. (After Vickery, 1973.)

Table 5.1 Values for several typhoons at Waglan Island, Hong Kong

Typhoon	*Distance (miles)*	$\hat{V}$	$\overline{V}$	K_τ	*Date*
Rose	13	52.3	39	1.34	16/08/71
Dot	7	29.4	20.3	1.45	17/07/73
Elsie	18	48.7	32.9	1.48	14/10/75
Agnes	27	31.2	24.2	1.29	18/07/78
Hope	14	54.9	40.1	1.37	02/08/79
Mac	24	41.0	33.9	1.21	23/09/79
Ellen	7	62.9	44.3	1.42	09/09/84

The factor z_0 is known as the roughness length, and typical values for this and zero-plane displacement (d) are given in Table 5.2.

It can be seen that, with these relationships, a vertical velocity profile can be constructed so long as the basic statistical description of the characteristics is known. In particular, it is usual to want to know what is the wind speed that will occur on average only once in a 50-year period. This wind speed is popularly known as the 'fifty-year wind'. This terminology is unfortunate, since it tends to lend a deterministic air to a random quantity. In

Table 5.2 Typical values for roughness length z_0 and zero-plane displacement d

Terrain description	z_0	d
City centres; forests	0.7	15 to 25
Small towns; suburbs; wooded country	0.3	5 to 10
Outskirts of small towns; villages; countryside with trees	0.1	0 to 2
Open level country	0.03	0
Level grass plains with isolated trees	0.01	0
Very rough sea; flat land areas	0.003	0
Snow-covered farmland; inland lakes	0.001	0

fact, this is a 'probability 0.02' wind. In other words, there is a 0.02 annual probability of occurrence. During any 50-year period, there is a roughly 63% chance of occurrence.

Gradient height is normally considered to occur at somewhere between 800 m and 2 km. In practice, it could be calculated from the following relationship:

$$V_g = \frac{1}{\beta}\frac{U_*}{f} \tag{5.39}$$

where U_* is the friction velocity, f is the Coriolis parameter and β is a constant (determined to be 6). Additionally U_* can be calculated from

$$U_* = \frac{V_{10}}{2.5\ \ln(10/z_0)} \tag{5.40}$$

The Coriolis coefficient is given by

$$f = 1.46 \times 10^{-4}\ \sin\Phi$$

where Φ is the latitude.

Thus for winds over the sea close to Hong Kong, gradient height should be

$$z_g = \frac{0.12 \times 44}{6 \times 0.55 \times 10^{-4} \times 2.5 \times \ln(75/0.001)} = 570\,\text{m} \tag{5.41}$$

5.1.4 Extreme wind statistics

The approach up to this point has defined a wind speed and its relationship to a gust maximum wind speed. It is now necessary to establish the values for these parameters. For the case of Hong Kong, typhoons represent the design case.

The movement of air over the surface of the planet has been studied in various places. In one particular study, the spectrum of horizontal gustiness was established over a very large frequency range. The Van der Hoven spectrum is often referred to and has become a standard. In Fig. 5.7 this spectrum is shown and it can be seen that there are essentially two parts to the spectrum. The lower-frequency part is associated with entire weather systems passing a point on the Earth's surface and it can be seen that the spectral peak is at approximately one cycle every four days. There is also a semi-diurnal peak associated with day and night. These peaks together with supposed peaks at an annual rate form the macrometeorological peak. To the right of the Van der Hoven spectrum is the micrometeorological peak. This is associated with turbulent energy and reaches a peak at about one cycle every 40 s. It is in this region that we are concerned with the response of structures to wind actions.

The distribution of wind speeds against time may be described by several different functions. It is normally assumed that the distribution of these wind speeds is best described by a Fisher–Tippett type of distribution. If the exponent of this is 2, then the distribution becomes a Weibull distribution and the cumulative distribution function can be described as

$$P_V = 1 - e^{-VK/C} \tag{5.42}$$

where K and C are the coefficients of the fitting process.

Appendix C takes data from the city of Rio de Janeiro and establishes the description of the extreme wind statistics using this methodology.

5.1.5 Statistics for the occurrence of strong winds

Moving from the meteorological to the wind-engineering domain, we encounter a field that draws heavily from both. The description of statistics about the return of strong winds normally gives as an end-product a description of a wind in terms of its probability of occurrence. To obtain this basic wind speed suitable for any particular location, analysis of the wind data collected over a number of years is normally effected. Such techniques have developed and are epitomized by the description by Cook and Prior (1987). Recording stations are normally composed of a series of anemometers that are run not only by government agencies. The data must be corrected to an 'effective height' before incorporation, normally as measures of mean wind speeds averaged over a suitable period such as one hour. The statistics of extremes may then be used to provide a description of the likelihood of the occurrence of a particular wind speed at some location.

This basic technique can be supplemented to produce more efficient estimates drawn from fewer data. Firstly, data relevant to storms can be accumulated rather than, for instance, yearly data. Secondly, the best linear unbiased estimator (Lieblein BLUE) can be used to give more emphasis to the data for shorter periods, thereby estimating the extremes more efficiently. The use of the Lieblein BLUE estimator on Hong Kong's wind data is shown in Appendix D. This technique gives greater emphasis to data from the main body of the set, and so the confidence in extrapolating to extreme values is higher.

In general terms, there seem to be some general principles that reduce confusion in the estimation of extremes. Firstly, it is better not to mix mechanisms. If strong winds are generated by cyclonic weather systems and by thunderstorms, separation of the two types of data reduces errors. A study of thunderstorm activity (Twisdale and Vickery, 1992) showed that, when the mechanisms were separated, then the winds generated by thunderstorms at the stations that are located far from the coast are underestimated if the other mechanisms are not separated before the analysis. The use of Fisher–Tippett type I distributions is theoretically applicable to only one mechanism at a time. The use of a Fisher–Tippett type II distribution to simulate mixed mechanisms is merely pragmatism without a theoretical basis.

The methodology for ascertaining extreme wind statistics is contained in Appendices C and D. The conventional methods are presented using data from the city of Rio de Janeiro, and the Lieblein BLUE estimator is used on data from Hong Kong.

There is also another very useful method for establishing wind speeds when a very strong storm has passed through an area, causing damage or severe damage. This method, which may be termed ‘reductionism’, is epitomized by the approach after Cyclone Tracy (Walker, 1975) or after Hurricane Hugo (Sparks *et al.*, 1992). In the latter, Sparks assembled data from several military bases and airports, mostly with anemometers at non-standard heights, to establish the probable area in which the strongest winds occurred. He then used the diaries of some of the local inhabitants to establish the most probable wind speeds. Sparks and his co-authors commented that the observation of sea height was of particular use since its rise was correlated with the occurrence of the strongest winds. He was able to ascertain that at Bulls Bay, South Carolina, the maximum gust was approximately 65 m/s, and the mean at the height of the storm was 38 m/s. Sparks and his co-authors have produced a list of types of damage correlated with gusts of various speeds in unprotected areas.

5.1.6 Extreme wind data for Hong Kong

In this section, Hong Kong is used as an example to show the assessment of data records with the intention of answering two questions:

1 At a chosen probability of occurrence, what is the most likely characteristic of the wind field?
2 What variations are there likely to be, for this event, around the Territory?

Two sources of data were used. Firstly, the extensive records of the Royal Observatory, Hong Kong (ROHK) were made available. The site at Waglan Island was selected as being a good measurement of the wind field, as it presents itself to the Territory before being affected by landfall.

The anemometer at Waglan Island is 70 m above sea level, but because of the influence of the island itself the wind characteristics measured are considered to be more appropriate to a height of 90 m. In practice, this correction makes only a small difference to the absolute estimates of wind speeds.

The data for the period 1953 to 1994 were analysed using the Lieblein BLUE (best linear unbiased estimator) technique. The advantage of this analysis is that it analyses storm data rather than yearly maxima. Additionally, it places greater emphasis on the more commonly measured windspeed characteristics. The result of the analysis for Hong Kong (at the Waglan Island site) is as shown in Table 5.3.

This analysis assumes that the entire process being analysed is invariant with time, and that statistics produced are representative of the series of events over all time. This is actually presumed not to be the case at present, in that a warming Earth has a non-linear effect on weather patterns. In the particular case being studied, a statistically significant difference for the paths of typhoons generated in the western North Pacific has been established (Jeary, 1991). This effect has not been taken into account for the purpose of this analysis. The net effect is to lessen the actual yearly probability of occurrence, and to result in a code that is more conservative.

The statistics for the return period are different for the mean and for the gust wind speeds. A regression analysis has been performed on the ROHK data. The basic expression for a conversion of return periods is expressed as follows:

$$A = \frac{k + \ln(R)}{5 + \ln(50)} \tag{5.43}$$

where R is the required return period, and A is the factor by which the fifty-year wind speed must be multiplied to give the wind speed for the R-year wind speed, and $k = 4.1$ for gust speeds and $k = 4.9$ for mean wind speeds.

The implication of having a different k value for gusts and for the mean is that the relationship between gusts and mean changes with increasing wind speed. Since the relationship between peak gust and mean is given by equation (5.34):

Table 5.3 Wind data for Waglan Island, Hong Kong

	50-year return period	*100-year return period*
Mean wind speed	46.3 m/s	50.3 m/s
Gust wind speed	64.7 m/s	70.2 m/s

$$\hat{V} = \overline{V}(1 + g_v I_u)$$

and g_v is considered to be a constant at a value of approximately 3.7, then implicitly I_u, the turbulence intensity, must decrease as the mean wind speed gets greater. This is a reflection of the actual situation, and measurements are discussed later in this chapter.

It is instructive to look at the measurements made at the time of the landfall for Typhoon Ellen, since that typhoon was the last to pass extremely close to the Territory. The 'eye' of Typhoon Ellen passed between Hong Kong and Macau on 9 September 1984. The area of greatest wind speeds was therefore within a few kilometres of the centre of Hong Kong. Figure 5.12 shows a representation of the wind-field over Hong Kong as measured at the ROHK's anemometer sites throughout the Territory. It is noteworthy that the generated wind speeds are quite similar to the 50-year return period statistics produced above. The measurements can therefore be used as a guide to the amount of sheltering likely to result from a 'probability 0.02' wind predominantly from a southeasterly direction.

The standard approach in wind engineering suffers from the fact that, in using a spectral approach, average values of several parameters occur, and

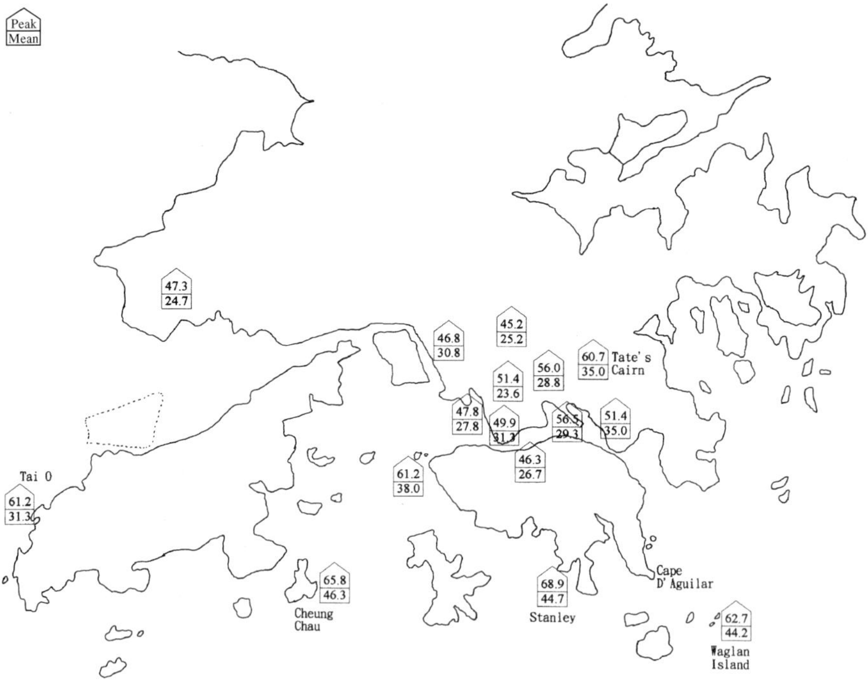

Fig. 5.12 Typhoon Ellen wind speeds (m/s) (figures shown are mean hourly and maximum gust). (Source: Oklahoma Geological Survey Observatory, USA.)

phase information is discarded. Average values are acceptable if the data being used are stationary. However, this is seldom the case when the wind is involved, and the departures from the average values can be relatively large. The time-history approach is a potential solution to this problem. Time histories are used in earthquake engineering and will be discussed below. However, the major design approach in earthquake engineering is also a spectrally based one.

The profiles for the extreme wind data vary with height according to the ground over which the wind is passing. Figure 5.13 shows the profiles for α values of 0.08 (to represent wind passing over the sea), 0.28 (to represent wind passing over the land) and 0.40 (to represent wind passing over a city). It should also be noted that the gradient height increases as a result of a greater ground friction coefficient.

In Fig. 5.14 the relationship between the $p = 0.02$ mean and gust wind for Hong Kong for general terrain (over land) is shown, whilst in Fig. 5.15 the relationship for the city of Hong Kong is shown. This result is constructed from measurements made during strong winds from various locations throughout the city. It should be emphasized that these characteristics are average characteristics. It is particularly interesting to note that the rationale for this construction is based on measurements of

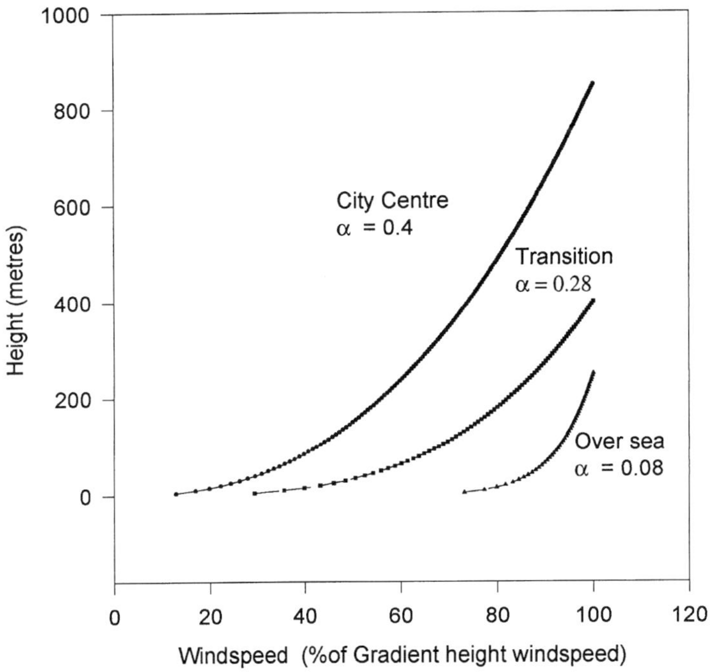

Fig. 5.13 Windspeed profiles over different terrains.

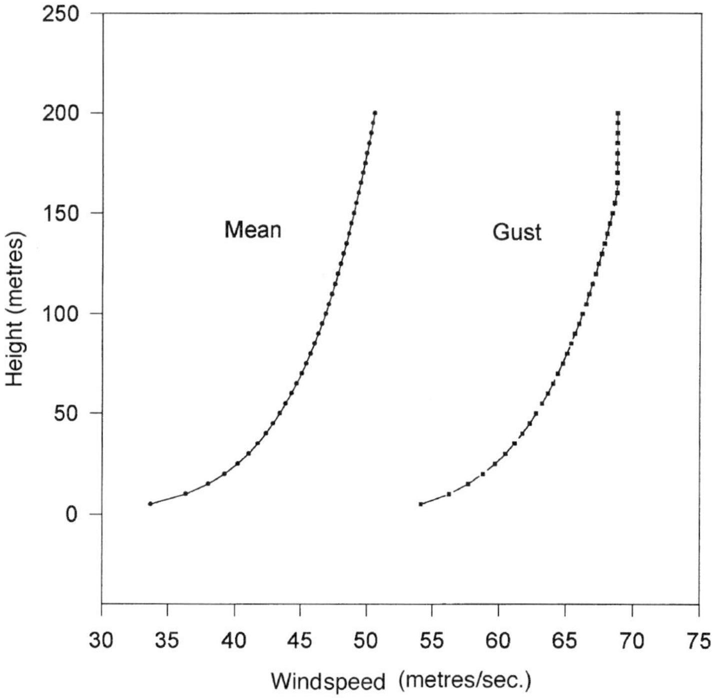

Fig. 5.14 Windspeed profiles for general terrain.

turbulence intensity at various heights, and that the profile for the maximum gust varies only in a small velocity range over a great height range. Indeed, this profile is perhaps the reason why some codes of practice assume that the effective gradient height reduces to as little as 150 m for tropical cyclones.

Figure 5.16 is a representation of the values of turbulence intensity at a height of about 350 m above the central business district of Hong Kong's urban area and over the sea. The values are assembled by making assumptions about the α values in both cases, that the values at gradient height must coincide, and that measurements over the city must be incorporated. Figure 5.16 is the result, and this characteristic was then used to construct the vertical profiles of mean and gust wind speeds for Hong Kong.

The interesting point about the measurements of turbulence intensity shown in Fig. 5.16 is that the difference in the intensity of turbulence, according to whether the path of the approaching wind is over the city or over the sea, is immediately apparent. The regression lines have been constructed assuming that Japanese data implying that an asymptotic value will be encountered above 20 m/s will apply to the measurement site as well.

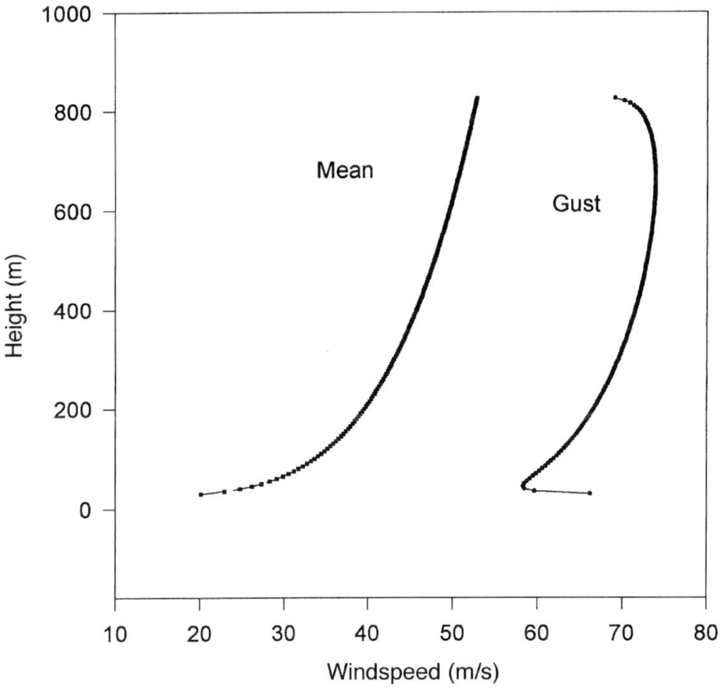

Fig. 5.15 Windspeed profiles for built-up terrain.

In Fig. 5.17 a comparison of profiles for general and built-up terrains is shown. At gradient height, all values converge as required.

The last full-scale typhoon to pass directly above the Territory of Hong Kong was Typhoon Ellen in September 1984. In fact, the 'eye' passed slightly to the west of Hong Kong. Figure 5.12 shows a map of Hong Kong and, superimposed on it, the measured values of mean and gust wind speeds around the Territory. The influence of mountains and of the gradual decrease in velocity with distance away from the 'eye' is clearly visible.

At this point it is important to emphasize the three-dimensional nature of weather systems. In the particular case of a village high in the mountains in Switzerland, for example, the village may well be situated above the weather of the valley below.

Figure 5.18 shows an Australian response to the lessons learned after Cyclone Tracy affected the city of Darwin on 25 December 1974. Because many roofs were stripped from buildings simply because they were not positively connected, the northern areas of Australia now require that positive connections are made. These connections are simple, and this particular example is taken from Nambour in southern Queensland.

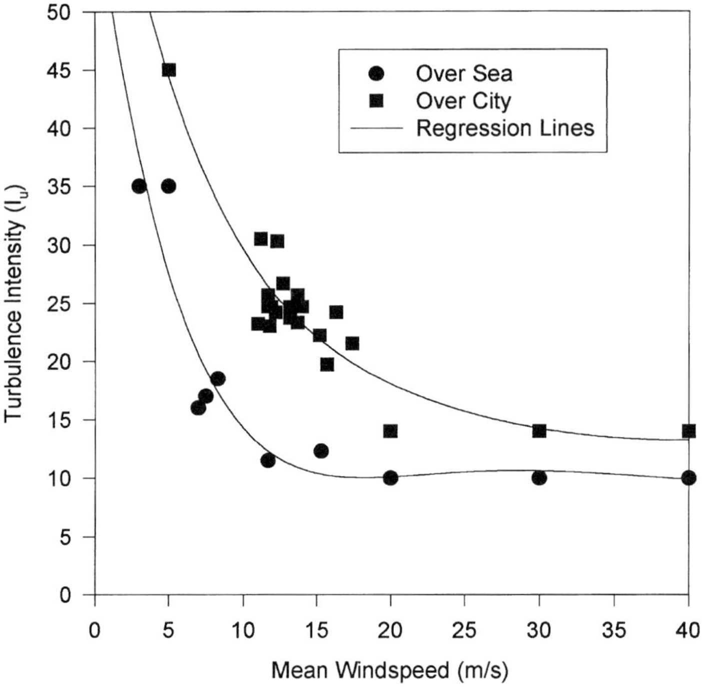

Fig. 5.16 Turbulence intensity, height 350 m, Hong Kong.

Figure 5.19 shows the awesome effect of a gale-force wind in England, in which a cooling tower is seen in the act of collapsing under a strong wind. Satellite dishes can also be destroyed by a strong wind. Such structures are very rarely designed with design-force winds in mind.

5.2 Earthquake engineering

A fully dynamic analysis of the response of structures to earthquake loading involves the assessment of the input action in terms of its spectral content or as an acceleration time history. Both of these approaches involve the assessment of either quantity at the site of the structure. With the use of spectra, the basic spectral approach previously outlined is used. If a time history is assembled, then a time stepping analysis will produce an estimate of the response time history.

These two basic approaches are, however, complicated by two important factors. Firstly, the relationship between the site characteristics and the source event are not only non-linear, but also, in any particular case, poorly researched. Secondly, the time between events is on a geological timescale rather than on a human timescale. We, however, are impatient to assess

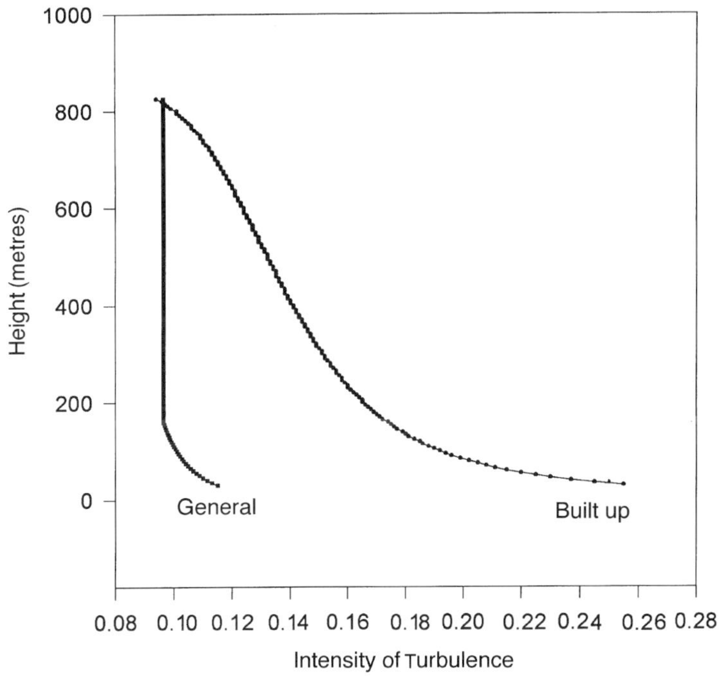

Fig. 5.17 Turbulence intensity profiles for built-up and general terrains.

Fig. 5.18 Part of Australia's response to the lessons of Cyclone Tracy, Nambour, Queensland.

Fig. 5.19 A cooling tower being blown down by wind action. (Source: Central Electricity Generating Board.)

such events, and are therefore induced into making predictions with insufficient information. Earthquake engineering, then, has to address difficulties associated with these two enormous problems, and has developed a number of techniques to try to minimize the effect of the unknowns. These two factors are considered here.

5.2.1 Relationship between source and site characteristics

The fundamental governing factor is the amount of energy released when a seismic event occurs. Such energy is available as a source and transmits its effects through the soil to the site being considered.

The nature and site of the seismic event dictate the quantity and the frequency content of the initial input. As this energy travels from the source to the site, this magnitude and frequency content are changed until they arrive at the site, and may be changed by the actual site conditions themselves. The magnitude and frequency content at the site may be assessed by either one of two basic methods.

In the former case, an identifiable source can be assessed, and the changes introduced in travelling to the site can be superimposed so as to arrive at an estimate of the site characteristics. Such an approach is normally assessed in terms of a site spectrum, although in theory a site time history could also be assembled.

In the latter case, measurements of response to earthquakes at the site (ideally) or at a similar type of site (or sites) can be used to assemble a site-specific spectrum or typical time history.

A time history is normally characterized in terms of the acceleration function. This is because of the simple linear relationship between the acceleration and the force acting on the structure.

5.2.2 Magnitudes and characteristics of earthquakes

The physical process causing earthquakes is known as plate tectonics. The process was discovered in the 1960s and was recognized after the work published by Barazangi and Dorman (1969). In this, they noted the positions of all earthquakes, and concluded that they were all confined to narrow belts, with only a few exceptions.

There are four major types of action:

1 Mid-ocean ridge activity causes expansion of plates. In this, stresses are small, and the lithosphere is thin, so energy release is quite small.
2 Two plates may pass tangentially to each other. This causes shallow earthquakes of large magnitude. The San Andreas Fault in California is of this type.
3 Subduction occurs at the boundary of an ocean plate and a continental one. The ocean plate passes under the continental one. The resulting earthquakes can be of any form, depth and magnitude.
4 Intra-continental earthquakes are brought about by intense compression causing brittle cracking. Typical of these are shallow-focus earthquakes occurring in mountain chains such as the Alps or the Himalayas. However, long-period events on seemingly flat plates are probably caused by the same mechanism. The events at Tenant Creek (Northern Territory, Australia) and New Madrid (Missouri, USA) are such instances.

On a worldwide basis, 90% or more of all earthquakes occur along or close to the plate boundaries. By monitoring the movement along such faults and knowing the average movement thereon, it is possible to know where stress relief occurs and where non-relieved areas are likely to induce large events at some time in the future. Nevertheless, it is difficult to predict with some certainty exactly when earthquakes will occur.

Having established the mechanism, it is possible to define the quantity of energy released in an earthquake by reference to the physical process. The seismic moment M_0 is defined as

$$M_0 = \mu S d \tag{5.44}$$

where μ is the shear strength of the material around the fault, S is the surface area of the fault and d is the average displacement of the fault. This leads to a definition of the magnitude of an earthquake as

$$M_w = (2/3)\log_{10}(M_0) - 10.7 \tag{5.45}$$

This definition is slightly different from the historically used definitions of magnitude, such as that introduced by Richter (1958). These definitions are based on measurements based on the analysis of earthquake-induced acceleration records. Figure 5.20 shows a record taken from the Northridge event in California. From such records, the arrival of P-wave and S-wave signals is normally easily observable.

Figure 5.21 shows one method of measuring ground motion. In this case an accelerometer is attached to a long spike driven into the soil, and responds to the motion of the soil. In this particular case, the soil motion was induced by using a vibrator system to shake the building in the background and monitoring the ground motion at progressively increasing distances.

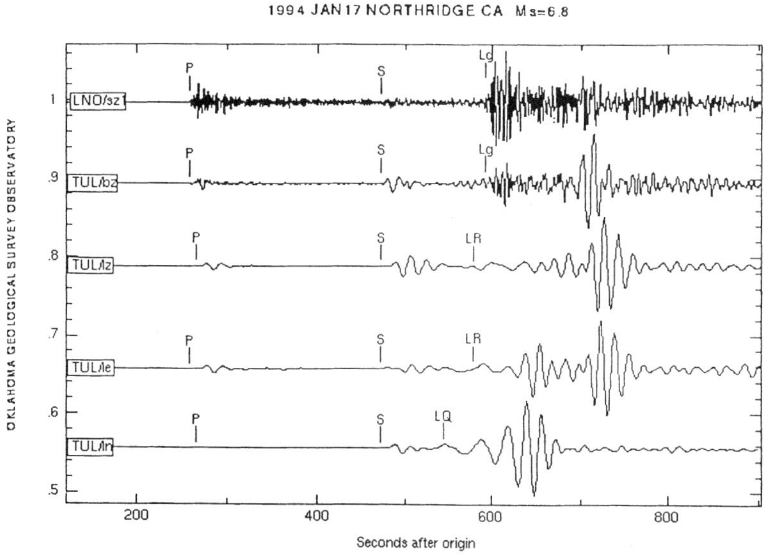

Fig. 5.20 Recordings of seismic waves from the Northridge earthquake. (Taken off the Internet. Source: Oklahoma Geological Survey Observatory.)

Fig. 5.21 Measurement of ground motion induced by a shaking building.

Using a record such as that in Fig. 5.20, it is possible to calculate the magnitude of an earthquake after the style of Richter. Richter's method only applied to California, and was modified in the following manner:

$$M_b = \log_{10}(A/T) + Q(D, h) \tag{5.46}$$

This is the body-wave magnitude formula, where A is the amplitude of ground motion (in micrometres, μm), T is the period of the vibration and $Q(D, h)$ is a correction factor based on D (the distance in degrees) and h (the three-dimensional distance from the epicentre to the measurement point). This can be modified to

$$M_s = \log_{10}(A/T) + 1.66 \log_{10}(D) + 3.30 \tag{5.47}$$

and this is known as the surface-wave formula.

The former of these formulae is based on measurement of the compressional P-wave signals, whilst the latter uses the Rayleigh surface waves.

The periods of the former are lower (being in the range of 1–5 s) whilst the latter are in the range of approximately 20 s.

In practice, each of these definitions from measurement (and all of the alternatives) give slightly different results because earthquakes do not produce the same proportions of frequency quantity in each case. In practice, larger earthquakes produce significantly larger proportions of energy at very low frequency.

The energy released by an earthquake can be related to the definition of M_s after the fashion of Gutenberg and Richter as

$$\log_{10} E = 11.8 + 1.5M_s \quad (5.48)$$

where E is in ergs. For every increase of 1 unit, the energy increases by a factor of 32. At the same time, an increase by 1 unit implies that the site amplitude increases by a factor of 10.

The last factor contributing to the input energy is the depth of the earthquake. In general terms, the deeper an earthquake, the greater the proportion of low-frequency energy. In addition, it is considered that the P-wave and S-wave signals are relatively uncomplicated when the earthquake centre is lower. The depth of an earthquake can be ascertained from the arrival times of different frequency components at remote sites.

The local intensity of an earthquake-generated action can be affected by local site conditions. Of particular importance is the situation of infill. In two recent cases, large amplification has occurred in such a circumstance. In Mexico City in 1985, infill of two old marshland areas created a condition in which a resonance of the landfill occurred when it was excited by an event of magnitude $M_s = 8.1$, some 300 km away at the coast. The resonance occurred at a frequency of approximately 0.5 Hz, and resulted in preferential damage to buildings in the 15 to 25 storey height range. (This is because the fundamental resonance frequency of buildings is a function of their height.) Estimates of the magnitude of the amplification have varied between 50 and 75 times, and the effect has been likened to parts of the ground 'ringing like a bell'.

In the second example, caused by the Northridge event in California, again infill occurred in certain areas and again caused an amplification of the ground motion under parts of the Nimmitz Highway. In this case the amplification by a factor of about 3 was enough to confound the local code predictions. Figures 5.22 and 5.23 show effects from the Northridge event. In the case of this particular building, the damage to one column looks quite severe, but when the entire building is viewed, then it can be seen that the damage is quite localized. In fact, much damage was of the nature of that shown in Fig. 5.24, in which cladding damage has been induced. The major damage that occurred in the Northridge event was to the Nimmitz Highway. Figure 5.25 shows the repair strategy taken, in which more and larger support columns were added to the structure.

Fig. 5.22 Damage to a column after the Northridge earthquake. (Courtesy of Dr John P.T. Mo.)

In Fig. 5.26 an instance of damage in the recent Kobe earthquake is recorded. In this instance, the earthquake occurred in an area that was considered not to have a significant risk of earthquake activity. The fact that the design did not consider the normal earthquake mechanisms led to this style of damage. In particular, the shear forces were simply too large for the structure to withstand.

The two largest recent earthquakes are those of 1960 in Chile, in which an energy release of 2.5×10^{30} dyn cm is reported, and the 1964 Alaska earthquake, in which 7.5×10^{29} dyn cm was released. For these two events, the magnitudes are reported in Table 5.4.

The energy release of the Chilean earthquake can be put into perspective by observing that it was, in broad terms, equivalent to an explosion of 200 megatons of TNT equivalent. To be able to cause structural damage, earthquakes need to achieve an energy release of about 1000 tons of TNT equivalent, and this corresponds approximately to a value of M_s of 5.5. The

Fig. 5.23 The building housing the column of Fig. 5.22. (Courtesy of Dr John P.T. Mo.)

Chilean earthquake represents the practical limit to energy release from an earthquake. In practice, larger energy releases would be possible but would surely be associated with events that would be described in other terms since they would have global implications.

Whilst these are the largest earthquakes recorded in modern times, with the benefit of scientific instruments to make a rational measurement of magnitude, it is also possible to obtain information from earlier earthquakes through the local descriptions of what occurred at the time. For this purpose, the modified Mercalli table of descriptions of effects is useful, and can be used to classify earthquakes in terms of the intensity of its effects. This table is reproduced as Appendix A.

Fig. 5.24 Damage to cladding from the Northridge earthquake. (Courtesy of Dr John P.T. Mo.)

Some other historical earthquakes and their estimated magnitudes are recorded in Table 5.5. In terms of M_s magnitudes, sometimes loosely described as the Richter magnitude, the general effects of earthquakes are as shown in Table 5.6.

5.2.3 Probability and location of events

It is estimated that in any given year the average number of 'major earthquakes' is 18 and there will be one 'great earthquake' somewhere in the world each year. The question often asked is whether there is an increase in the number of occurrences.

Figure 5.27 shows the number of major earthquakes that have occurred worldwide in the recent past. The average number since the turn of the century is 18, and it can be clearly seen that we are in a 'quiet period', as all but three years have a number below this average figure. It is also apparent that the numbers have been dropping throughout the last 30 years. A run

Fig. 5.25 Repairs to the Nimmitz Highway after the Northridge earthquake. (Courtesy of Dr John P.T. Mo.)

Fig. 5.26 Damage at a joint after the Kobe earthquake. (Courtesy of Dr Yukio Tamura.)

Table 5.4 Magnitudes of the two recent largest earthquakes

	1960 Chile	*1964 Alaska*
Energy release (dyn cm)	2.5×10^{30}	7.5×10^{29}
M_s	8.5	8.3
M_w	9.6	9.2

Table 5.5 Other earthquakes and their estimated magnitudes

Location	*Date*	*Magnitude* M_s
Lisbon, Portugal	1 November 1755	8.7
San Francisco, USA	1906	8.0
Messina, Italy	28 December 1908	7.5
Gwanxu, China	16 December 1920	8.6
Kwanto, Japan	1 September 1923	8.3
Xining, China	22 May 1927	8.3
Gwanxu, China	25 December 1932	7.6
Quetta, India	30 May 1935	7.5
Peru	31 May 1970	7.8
Tangshan, China	27 July 1976	8.0
Iran	20 June 1990	7.7

Table 5.6 Correlation of effect with magnitude

Magnitude M_s	*Effect*
3.5	Threshold of human perception
3.5 to 5.4	Perceived by humans
5.5 to 6.0	Minor damage to structures
6.1 to 6.9	Can be destructive depending on soil conditions, and construction style
7.0 to 7.9	Major earthquake; creates major damage to structures with exceptions only for the best anti-seismic design
8.0 or greater	Great earthquake; destruction of structures

test for stationarity establishes a bias indicating a reducing trend at the 99% confidence level.

Some 90% of earthquakes occur along or close to plate boundaries; however, this and other statistics only go some way to establishing the probability and type of loading that would occur at any site.

Probability of occurrence can be established for any particular site if records are sufficiently detailed. Short return periods of the order of several decades are possible to establish with modern records. However, return periods of the order of centuries are very difficult to establish because of the dearth of data. For any particular site, statistics for average return period

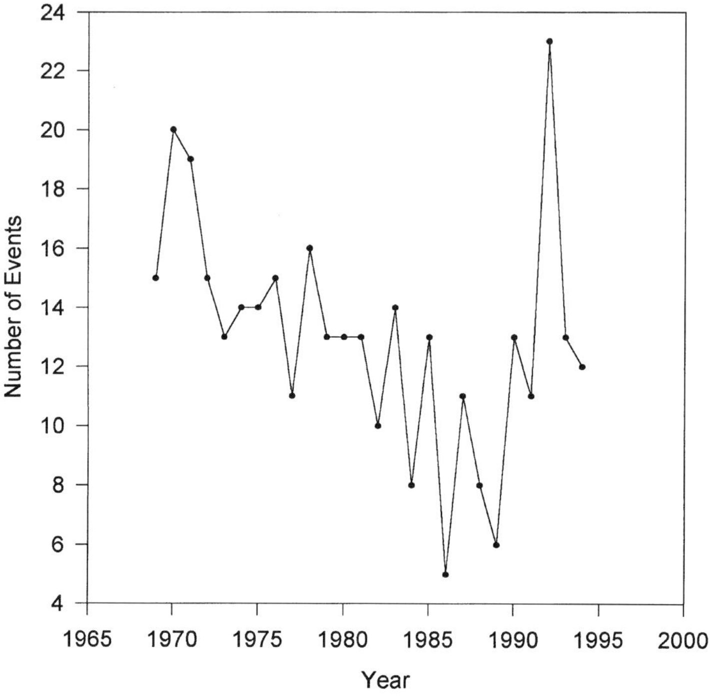

Fig. 5.27 Worldwide number of major earthquakes, 1969 to 1994.

can be established, and from that the probability of occurrence can be determined.

The form of the spectrum of earthquake action is shown in Fig. 5.28. This is a generalized form and there has been controversy about this form as well. In particular, the value of f_{max} is open to question, since it may itself be a function of the soil conditions along the path, and may therefore be modified at different measuring stations given the same earthquake.

The point here is that the designer must attempt to get an estimate of the input function at his or her site. A spectral form that gives an amplitude and a frequency content is clearly helpful. The accuracy of the estimate is another matter. It is clear that information from a site, or at the very least close to it, is essential if a good estimate of response is to be obtained.

An alternative to the spectral approach is to estimate a time history at a site. The same problems as to frequency content will still exist, however, and, once again, the more specific the information is to the site, the more precise will be the estimate.

In practice, the response of a structure to an earthquake loading is more a function of the excitation rather than the structural characteristics. This is the case only if the structure remains monolithic. It is therefore quite

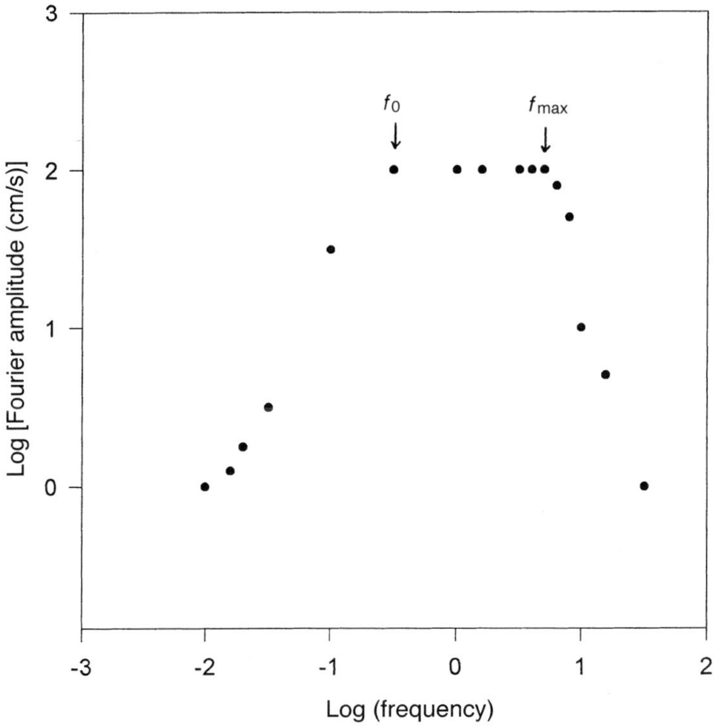

Fig. 5.28 Spectrum of earthquake action.

common to see response spectra as a method of estimating the response of a structure to an earthquake action. Figure 5.29 gives an example of this approach. There are different response curves for different damping values. Under earthquake loading, structures very quickly move to use a maximum of available damping, and the values tend to be larger than is the case for wind loading. In practice, the response with a structure that has a damping value of more than 5% tends to a very similar value.

It is tempting to try to introduce a relationship between the modified Mercalli (MM) intensity and the maximum acceleration experienced at a site. Unfortunately, the correlation is rather poor, and for any particular MM intensity the acceleration tends to extend over an order of magnitude or more. The mean of the relationship is

$$\log a = 0.25 I_{MM} + 0.25 \tag{5.49}$$

where a is in cm/s^2 and I_{MM} is the modified Mercalli intensity. Thus for MM VIII scale, the predicted value would be 18% g whilst measurements of acceleration have ranged from 8 to 70% g.

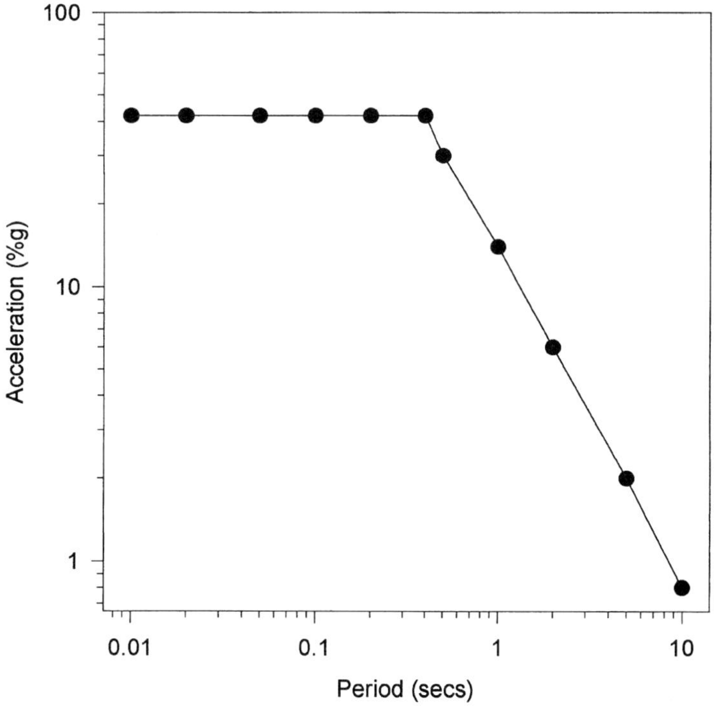

Fig. 5.29 Design spectrum for 5% damping.

A more reliable estimate of the duration of the earthquake against Richter magnitude is shown in Fig. 5.30.

5.2.4 Design earthquakes

Several widely different methods exist for establishing the first part of the systems equation. This is the forcing function. In practice, this is sometimes combined with the $H(f)$ function as well in earthquake engineering.

Equivalent quasi-static loading in codes

The value of these methods is that they are relatively simple, and they allow the use of algebraic equations for the calculation of a response that purports to take into account the dynamic response. Factors for intensity, ground type and building characteristics are normally included.

In all earthquake codes, the basic approach is similar. There is a requirement that the lateral load applied is

$$v = Kw \tag{5.50}$$

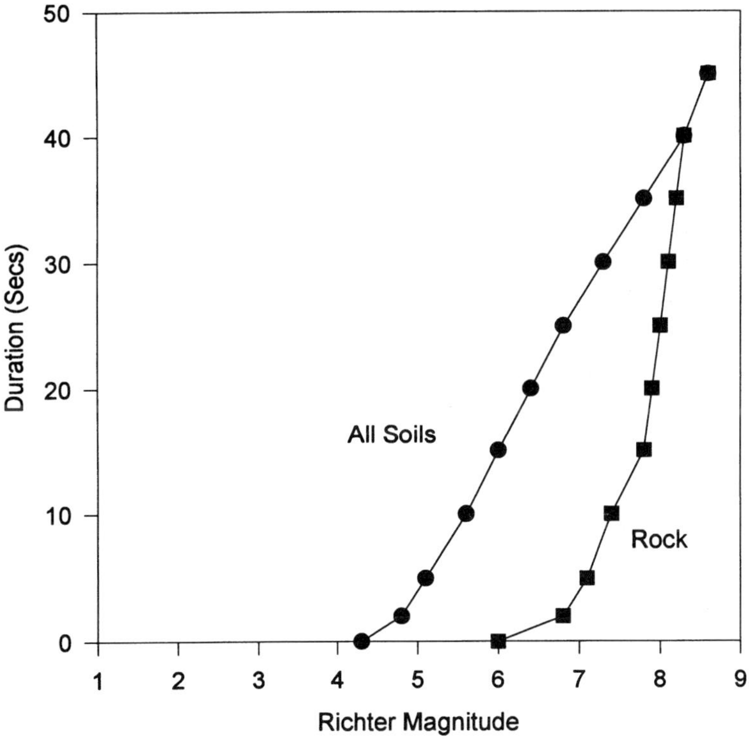

Fig. 5.30 Duration of earthquakes.

where w is the dead weight of the building, K is a constant depending on soil and structural parameters and v is the lateral force, distributed over the height of the building.

Different countries use different values of K to account for their own earthquake regime and their soil conditions, and this factor is often separated into coefficients to account for the separate influences.

This is implemented as follows in the New York City Seismic Code (NCEER, 1991), in which the horizontal base shear for design purposes is taken as

$$V = \frac{ZIC}{R_w} W \tag{5.51}$$

where Z is a factor dependent on the zone of construction, and is based on a chosen return period, I is a multiplying factor to account for the importance of the structure, R_w is a factor to account for the type of design (e.g. shear wall, moment frame, etc.),

$$C = S/(T^{1.1})$$

T is the fundamental period (= the inverse of the frequency) and S is a soil factor. Values of R_W can be between 3 and 6. For a shear-wall type of construction, it is 6, whilst for a moment-resisting-frame type of design, it is 4.

The values calculated in this way can be shown as a diagram of acceleration against period, as in Fig. 5.29. The approach to design can then be taken as a quasi-static equivalent, in which forces are distributed against height of the structure using the influence function approach. The New York Code (NCEER, 1991) takes a straight-line mode as the recommendation. The code approach is intended to arrive at a similar prediction of response as that which would be obtained if a more rigorous methodology was used.

Response spectra

Response spectra can be assembled for a site. These are produced from measurements of design events only if the designer is lucky enough to have the benefit of a measurement at the site or in the near region. More likely is that such a spectrum will be an amalgam of information sources from events that are assumed to be typical of what might occur in the design event at the site. The approach in California (SEAOC, 1988) and in the Uniform Building Code of the USA is to assemble data about events that have a return period of 475 years. The basic ground acceleration spectra then have characteristics imposed on them to account for different damping values.

Accelerograms

In much the same way as design spectra are assembled, then accelerograms may be constructed from site or similar data. These records ideally come from an averaging or collection of similar events at a similar site. However, it is also possible to produce simulations of acceleration time histories. To effect this, a computer simulation of white noise (random signals having energy at all frequencies) is the starting point. Actually the white noise is 'band-limited' white noise, since it is impossible to have an infinite bandwidth, and so it should really be called 'pink noise'.

This noise signal is then shaped according to the required spectral characteristics, by passing through a filter with the desired characteristics. Characteristics such as those of Fig. 5.29 would be chosen for this purpose. The resultant is a series of random events, all of which have the same spectral characteristics, but completely different time history sequences. In this way, a structural response can be tested against such potential inputs.

The purpose of such an approach is to take account of the phase information. The time of arrival of the various components of the acceleration time history has a marked effect on the structural response. This is caused by the fact that, if the arriving force is in phase with the response, it will enhance it, whilst if it is antiphase it will oppose the response and reduce it. The situation is analogous to walking towards a door that is swinging open sequentially on both sides of the closed position. If, as you walk towards it, it is moving away from you, you can pass through with little effort. However, if it is swinging towards you, your passage will be more difficult. The energy of swinging of the door is the same in both situations, but the phase of the swing is all-important to the ease of your passage.

5.3 The Duhamel integral

The difficulties associated with spectral techniques have been noted above. These are caused by the abandonment of phase information when using spectral techniques. The solution to the problem could be arranged by using a complex frequency response function for the 'system response', and couple this with complex spectra for the forcing action. However, a rather simpler approach can be arranged when operating in the time domain.

In the time domain, the force action and the system response are both shown as a function that varies with time. The attraction of such a methodology is that it is relatively straightforward to calculate responses even when the action is severely non-linear or non-stationary. There are several methods for calculating a response time history. The Duhamel integral is one such method.

The basic approach of the Duhamel integral is to assume that, at a finite time τ, a loading action will cause a response of a system during a time greater than τ. This response is termed the response to an impulse function. The entire load action can then be represented by a series of impulse functions, and the response to all of them is summed together to produce the complete response of the system.

This process can be represented as follows:

$$R(t) = e^{-\zeta\omega(t-\tau)} \frac{F(t)\,d\tau}{M\omega_n\sqrt{(1-\zeta^2)}} \sin(\omega_n) \sqrt{(1-\zeta^2)}\,(t-\tau) \qquad (5.52)$$

If we sum these response time histories over the time of the load action, then we obtain

$$R(t) = \frac{1}{M\omega_n\sqrt{(1-\zeta^2)}} \int_0^t e^{-\zeta\omega(t-\tau)} F(t) \sin(\omega_n) \sqrt{(1-\zeta^2)}\,(t-\tau)\,d\tau\, h \qquad (5.53)$$

and the response of the system to a unit impulse is

$$h(t-\tau) = \frac{1}{M\omega_n\sqrt{(1-\zeta^2)}} e^{-\zeta\omega(t-\tau)} F(t) \sin(\omega_n) \sqrt{(1-\zeta^2)}\,(t-\tau) \tag{5.54}$$

For a numerical solution to this equation, it can be rewritten as

$$R(t) = A(t) \sin\left[\omega_n\sqrt{(1-\zeta^2)}\,t\right] - B(t) \cos\left[\omega_n\sqrt{(1-\zeta^2)}\,t\right] \tag{5.55}$$

where

$$A(t) = \frac{1}{m\omega_n\sqrt{(1-\zeta^2)}} \int_0^t F(t) \frac{e^{-\zeta\omega\tau}}{e^{-\zeta\omega t}} \cos\left[\omega_n\sqrt{(1-\zeta^2)}\,\tau\right] d\tau \tag{5.56}$$

and

$$B(t) = \frac{1}{m\omega_n\sqrt{(1-\zeta^2)}} \int_0^t F(t) \frac{e^{-\zeta\omega\tau}}{e^{-\zeta\omega t}} \sin\left[\omega_n\sqrt{(1-\zeta^2)}\,\tau\right] d\tau \tag{5.57}$$

The evaluation of the first integral using Simpson's rule is

$$A(t) = \frac{\Delta\tau}{m\omega_n\sqrt{(1-\zeta^2)}} \frac{1}{3} \sum_3^A (t) \tag{5.58}$$

where

$$\begin{aligned}\sum_3^A (t) = \sum_3^A \Big\{ &(t-2\Delta\tau) + F(t-2\Delta\tau) \cos\left[\omega_n\sqrt{(1-\zeta^2)}\right](t-2\Delta\tau)\, e^{-\zeta\omega 2\Delta\tau} \\ &+ 4F(t-\Delta\tau) \cos\left[\omega_n\sqrt{(1-\zeta^2)}\right](t-\Delta\tau)\, e^{-\zeta\omega\Delta\tau} + F(t) \cos\left[\omega_n\sqrt{(1-\zeta^2)}\right] t \Big\}\end{aligned} \tag{5.59}$$

The $B(t)$ term is similar but involves sine terms.

Figure 5.31 shows the result of a calculation using the Duhamel integral. The upper trace is a real record of wind speed, and this is converted to a force to represent $F(t)$. The lower trace is the calculated response to this forcing function for a structure with a natural frequency of 1 Hz and 1% damping.

5.4 Sinusoidal excitation

If the excitation source is sinusoidal, then the principle of simple harmonic motion can be applied absolutely. Under such a circumstance, the relationships between displacement, velocity and acceleration are well defined.

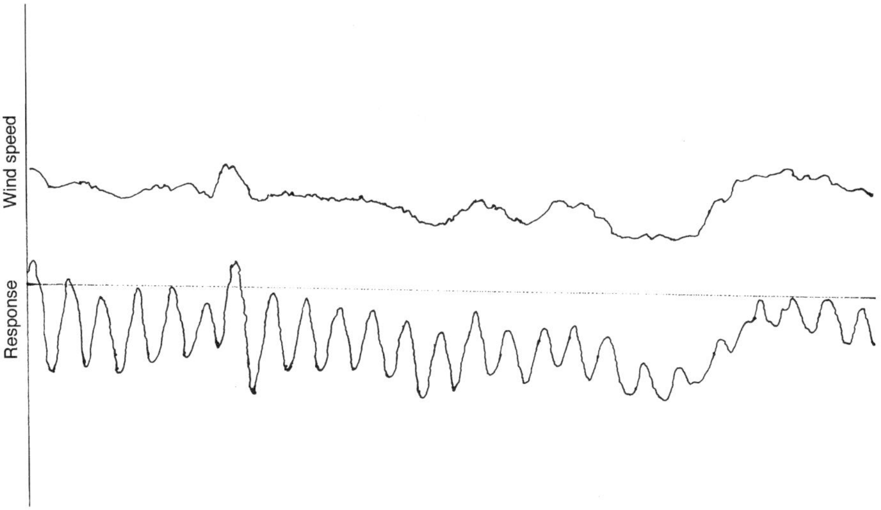

Fig. 5.31 Response of a structure calculated using the Duhamel integral.

Equations (3.27) and (3.28) defined motion at resonance. They are simple algebraic equations and show that, if four of the five parameters (mass, force, damping, displacement, frequency) are known, then the fifth can be calculated. This principle is used in induced vibration testing to establish the entire mass of a structure.

Additionally, by varying the frequency, controlling the force (induced by a very large vibrator system) and measuring the acceleration of the structure at each frequency, it is possible to establish the complex frequency response function $H(\omega)$ of the structure, since effectively the force input and the response are measured. The complex frequency response function is then merely the response per unit force as a function of frequency.

Because the principle of simple harmonic motion is invoked, then the following relationship for converting a measurement of acceleration to a displacement holds:

$$248.5g/f^2 = a \qquad (5.60)$$

where g is the acceleration (expressed as a fraction of g), f is the frequency (in Hz) and a is the displacement (in mm).

Figures 5.32 and 5.33 show two different vibrator systems used by the author to shake entire structures. In Fig. 5.32 a synchronized four-unit rotatory system is shown deployed inside the 200 m tall Emosson Dam in Switzerland; whilst in Fig. 5.33 a single reciprocating unit is deployed on the top floor of a 35-storey building in Hong Kong.

Fig. 5.32 A four-unit synchronized shaker system inside Emosson Dam.

Fig. 5.33 A reciprocating shaker deployed in a tall building.

5.5 *Other types of loading*

By now, it should be clear that the systems approach allows us to describe actions and response under any circumstances. Loadings caused by actions such as gas explosions, nearby piling or blasting operations, accidental impact by vehicles and human-induced vibration all allow the designer to gather information about the force input, and then to use this to interact with the influence function and the representation of the structure to produce a representation of the response of the structure in terms of its overall magnitude of response and of its deflected form.

Of interest in this respect is the example of crowds of humans joining together in frenzied 'pogo dancing', in which synchronized jumping takes place.

CHAPTER 6 *Structural performance*

In this chapter, the response of structures to actual forces is considered. However, here the approach is not the conventional one in which a theoretical response is considered, but rather the response of real structures to real forces is considered and discussed.

It is necessary in this chapter to invite the reader to 'think dynamic'. The conceptual understanding of the dynamic behaviour of structures can best be achieved if the dynamic response is visualized as a series of modal responses. For the sake of simplicity, it is probably best to start with the 'mind's eye concept' of a single mode of vibration, or of a group of fundamental modes of vibration. Having achieved an understanding of the 'normal' modes listed previously in this book, it is necessary to introduce a further concept. This is that each structure is individual; it behaves differently from the normal. Sometimes this difference is so small that it seems as though the behaviour is completely correlated with the theoretical models; however, sometimes the differences are quite large.

The series of full-scale induced vibration tests referred to in this text alerted this author to the fact that, once such testing became a routine, the structure about to be tested was almost certain to confound the investigators. As soon as the team believed that it truly understood dynamic behaviour and was merely confirming that with field tests, it was almost certain that the structure would introduce a difference from the expectation.

What follows is a discussion, by structural type, of some of the issues that have become apparent from the testing of full-scale structures. Those tests were initially performed using large shakers that induced motion and allowed the investigation of a steady-state dynamic response. As techniques improved, then more information from random excitation became usable.

For the purposes of clarification, structures are considered in turn by type.

6.1 Tall buildings

Most of the dynamic response of tall buildings occurs in the fundamental modes of vibration. There are normally three such modes, comprising two orthogonally oriented translational modes and a torsional one.

Figure 6.1 shows the 'normal' form of such a triad of modes of vibration, represented in the form of a spectral response. This particular example is taken from the University of Sheffield's Arts Tower in the UK (Fig. 6.2). The spectrum is an ensemble average in which very short-term spectra are archived, and then the ensemble average is performed only over those spectra occurring when the wind blew from a certain direction. In this particular case, the direction, shown on the figure, is at a mean angle of 280°.

The response can be seen to be predominantly north–south (the long axis of the building), with a much smaller response in the orthogonal direction. The torsional response is relatively large because the torsional moment is also large. An ensemble average like this one can only be established when a great deal of data are archived. In this particular case, this result was achieved only after collecting data for more than two years.

In such a conventional case, then, the principal issue concerns the frequency juxtapositioning of the three modes. Again, the norm can be considered to be that, in the frequency domain, the torsional mode of vibration occurs between the two translational ones. The frequency difference between the two fundamental modes is normally 10% of the lowest frequency. The factors referred to above occur because of the following of standard conventions in the design of tall buildings. Where such

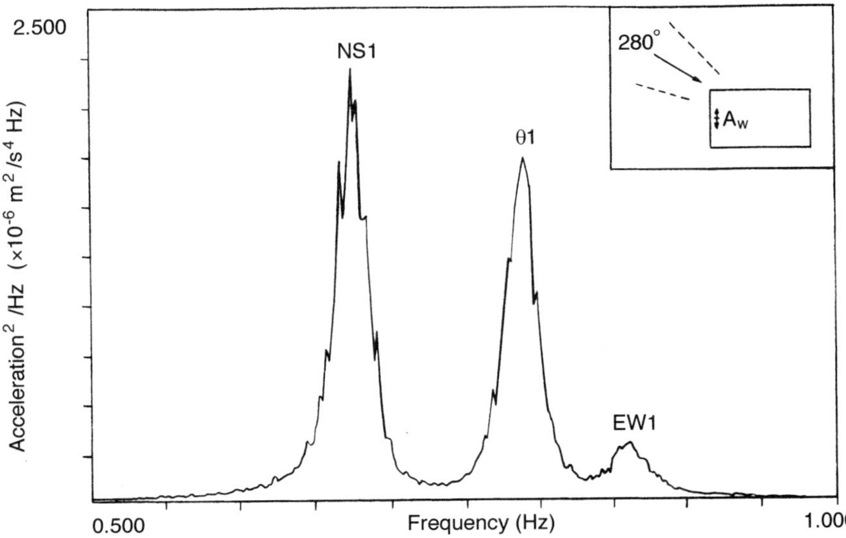

Fig. 6.1 Ensemble-averaged spectrum from the University of Sheffield Arts Tower. The spectrum is an average of 615 records. Mean wind speed was between 10.0 and 15.0 m/s, with an average of 12.2 m/s. Wind direction was between 265° and 295°, with an average of 280°.

Fig. 6.2 The Arts Tower at the University of Sheffield, UK.

conventions are not followed, then the differences from the norms become larger.

The norms may be considered to be buildings with an aspect ratio of close to 1:1, and with a conventional distribution of shear elements. In other words, there may be a core to the building, and columns forming, or close to, the facade of the building. Such an arrangement of stiffness elements dictates the torsional stiffness and dictates the relationship of the torsional frequency with the translational ones.

The lowest-frequency translational mode of vibration will occur in the direction in which the translational stiffness is the least, and the orthogonal fundamental mode occurs at 90° to this. The direction of least stiffness aligns with the geometry of the building for buildings that are of a symmetrical shape. However, in cases where the building is not symmetrical or where there are significant appendages to the main structure, then the directionality of the modes may well differ from the geometry of the structure itself. The dominant factor, though, is that the direction of least stiffness dominates. This directionality may not always be obvious. Notable examples are shown below.

The National Westminster Tower in London (Fig. 6.3) has a modal directionality that would be difficult to predict from a visual inspection, because the three arms of the building do not span over the same number of floors. Figure 6.4 gives a representation of the plan of the building, with the modal directionality superimposed.

In the case of the Leicester University Engineering Block (Figs 6.5 and 6.6), the appendage forms such a significant part of the structure that the directionality is not obvious. The modal directionality shown in Fig. 6.7 was obtained from experiment. This experiment was simply performed by shaking the building, and taking motion readings at 5° intervals through a 180° arc. The minimum response then establishes the modal directionality, as shown in the figure. In the case of the Leicester University Engineering Block, the appendage was of such significant proportions that two torsional modes of vibration appeared. The first was a conventional rotational type, whilst the second was more reminiscent of the type of dynamic movement undergone by a set of weights, when lifted by a weightlifter, with the two end masses swinging about the connecting bar. Such imaginable analogies

Fig. 6.3 The National Westminster Bank Tower, London, UK.

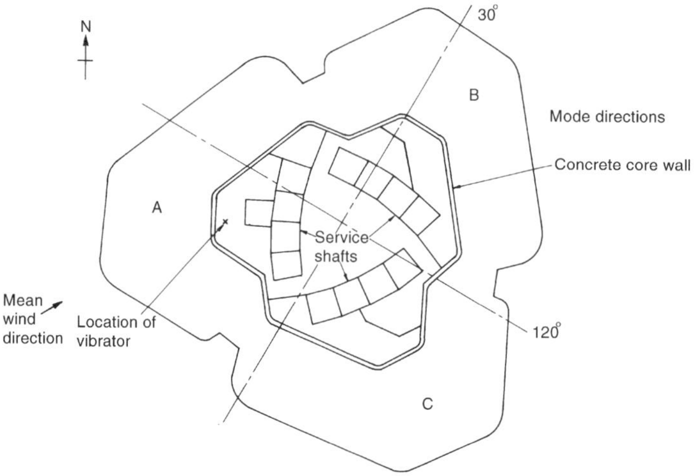

Fig. 6.4 Modal directionality of the NatWest Tower, London, UK.

Fig. 6.5 Leicester University Engineering Block and Tower.

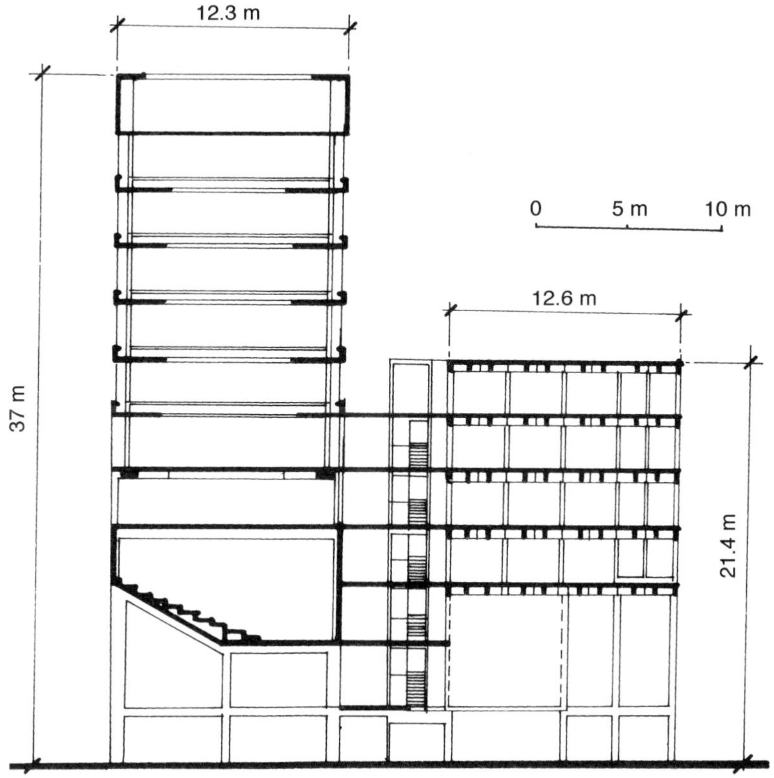

Fig. 6.6 Section of Leicester University Engineering Block.

are particularly helpful in the process of 'thinking dynamics'. Figure 6.8 shows a representation of the modes of vibration measured at this Leicester University building.

It is important to emphasize the point that appendages affect structural behaviour. In the case of the Cathedral at Chartres, for example, the flying buttresses are particularly important in preventing buckling of the walls caused by the weight of the roof.

It is possible to conceive of other shapes of buildings in which the directionality is changed from the geometric alignment of the building. Two common forms of structure are of particular relevance here. In the first an 'L' form changes the directionality, and in such a case the positioning of the stiffness elements is of fundamental importance. The second common type of tall building involves an essentially circular cross-section. In this type of case, the modal directionality is established by quite small differences in the stiffness around the 360° arc. It is normally the core of the building that establishes modal directionality in such a case. Torsional modes for current circular buildings are normally higher and well separated from the two

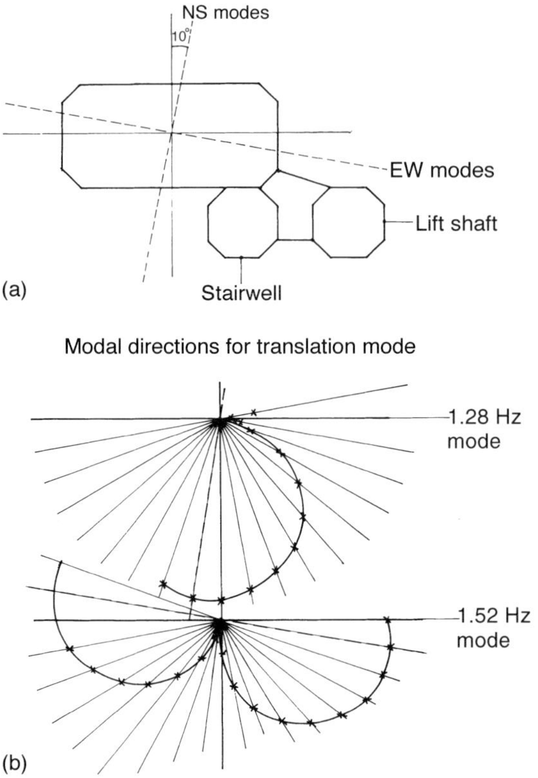

Fig. 6.7 Directionality test at Leicester University Engineering Tower.

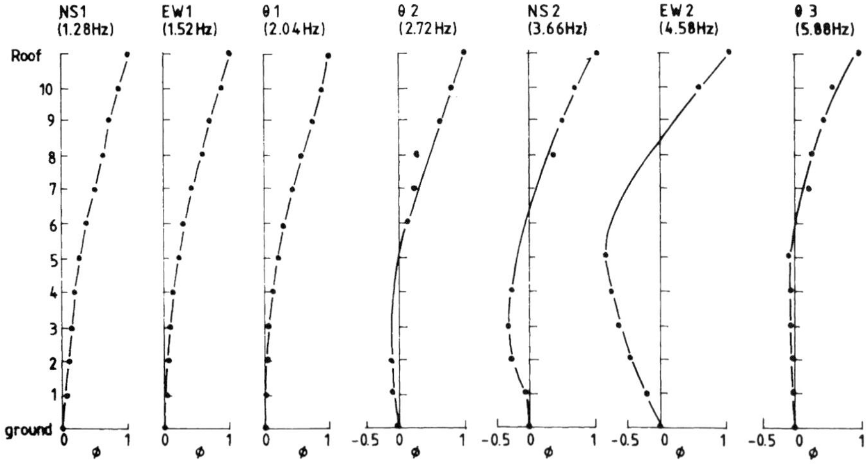

Fig. 6.8 Leicester University Engineering Tower mode shapes.

translational modes of vibration, because it is normal to concentrate many columns around the periphery. The periphery is in any case not normally a large distance from the core.

The next question of importance is that of modal coupling. In the case of tall buildings, the situation is relatively uncomplicated, although the problem can be significant. Modal coupling occurs when vibrational energy is swapped from one mode of vibration to another. The medium for such coupling can be through the fundamental torsional mode, or it may be caused by geometric coupling.

In the former case, in which energy is swapped into the torsional mode and then into the translational orthogonal mode, it is sufficient that the natural frequencies of the torsional and translational modes are close. That such a situation occurs is witnessed by Fig. 6.1, in which the frequency spectrum monitored at the University of Sheffield Arts Tower was observed to have all of the three fundamental modes within a frequency span of only one-quarter of a cycle per second. In such a case, energy is easily swapped from one mode to another when there is a random loading source. This behaviour is observable from the diagrams showing the modal displacements (Fig. 6.9). In the Y–Y direction (north–south direction), coupling with the torsional mode is clearly observable.

In the case of a geometric coupling, the necessary precondition is that the centre of torsion is offset from the geometric centre of the building. In such a case, a load appearing on one face of a building, even if evenly distributed, will produce a response both in translation and in torsion. Figure 6.10 demonstrates the point. This form of coupling is seen in practice in Fig. 6.11, in which the open-plan St Peter's House building (Oldham, UK) was constructed with very few stiffening elements, and with a lift shaft with four shear walls near to one end of the building. The resultant coupling in the X–X direction is not surprising.

Torsion–translation coupling is inconvenient for design since it is inevitable that, where the mechanism exists, then all of the vibrational energy will, sooner or later, appear in one single mode of the triad. In such a circumstance, the response will be significantly larger than conventional calculation would have us believe. The design approach should be firstly to try to avoid the situation; but if this is not possible, then the calculation of response must take a 'worst-case' view of the behaviour. This is achieved by summing the energy in the three modes under a design random loading and increasing the response of each mode of vibration in direct proportion.

When the form of construction includes more stiffening elements, such as in the case of the Civic Centre, Plymouth, UK, shown in Fig. 6.12, it can be seen that no measurable coupling takes place, despite the fact that the three frequencies are close together.

A further interesting feature of the response of the Civic Centre is that in mode $Y–Y_1$ the fundamental mode appears in a form more reminiscent of a

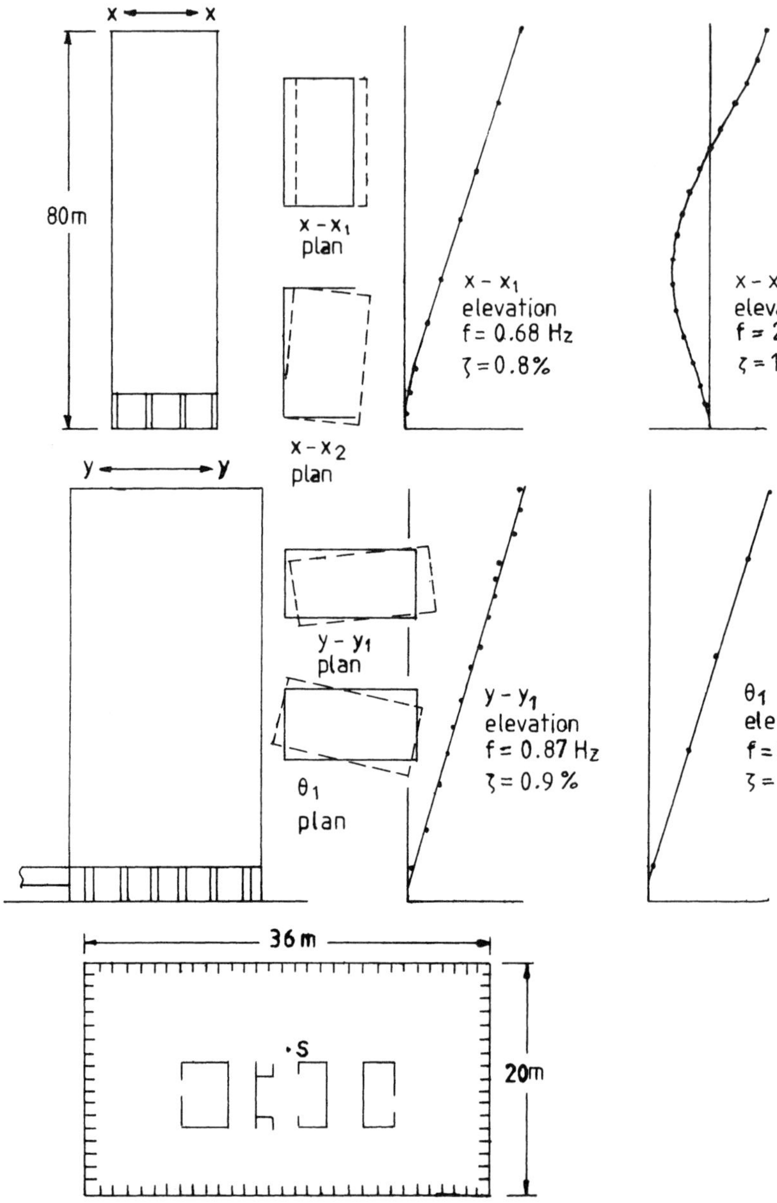

Fig. 6.9 Mode shapes for the Arts Tower, Sheffield University.

shear-type mode. The large mezzanine level undoubtedly contributes to the two fundamental translational modes' lack of movement at these levels, but it does not explain the shear-type deflection. It is also worth noting that, in

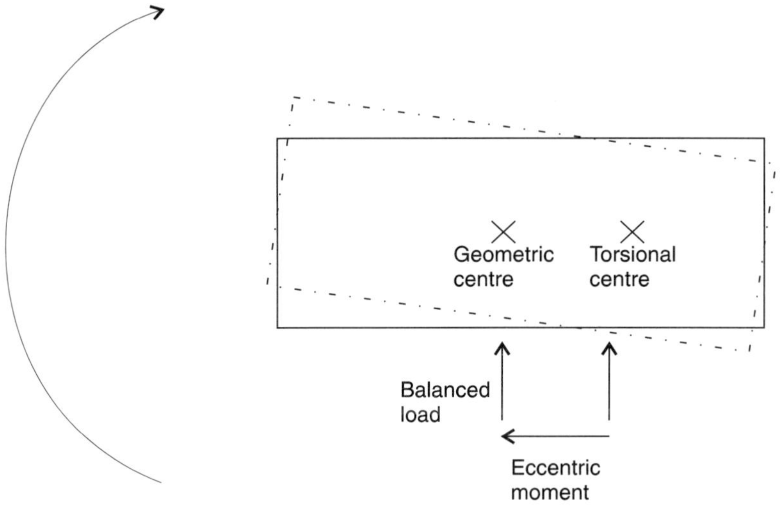

Fig. 6.10 Geometrically induced modal coupling.

the second translational and torsional modes, displacements pass through the mezzanine floor and continue into the basement.

A similar shear-type fundamental mode appears for the $Y–Y_1$ direction of the British Rail building at Plymouth, shown in Fig. 6.13. Figure 6.14 shows a comparison of the measured modal deflection with the theoretical shear deflection. The similarity is quite striking. Such instances are the exception rather than the rule, and Figs 6.15 and 6.16 are rather more typical of mode shapes for tall buildings in general.

In Fig. 6.17 a comparison of the modal deflection for the second translational mode measured at Sheffield University Arts Tower is compared with the theoretical curve for such a mode. In this case, the deviation is rather greater, and this tends to be a feature of second translational modes. They tend to show more variations from the theoretical curves. This is caused by the fact that, since the distortions in the structure are greater, irregularities or variations to stiffnesses have a proportionately greater effect.

The last factor to consider for tall buildings is that the response to a dynamic loading is a function of the modal stiffness in each case. The modal stiffness itself is a function not only of the stiffness elements in the building, but also of the stiffness provided by the foundation–soil system. Once again, there may be surprises in evaluating this aspect. A test on a building close to a tidal river produced a variation of the fundamental natural frequency. Subsequent investigation showed that this effect was correlated with the tides in the river, but with an 8 h delay. The building was constructed on a clay soil, and the repeated saturation and dehydration of the clay caused a variation of the stiffness.

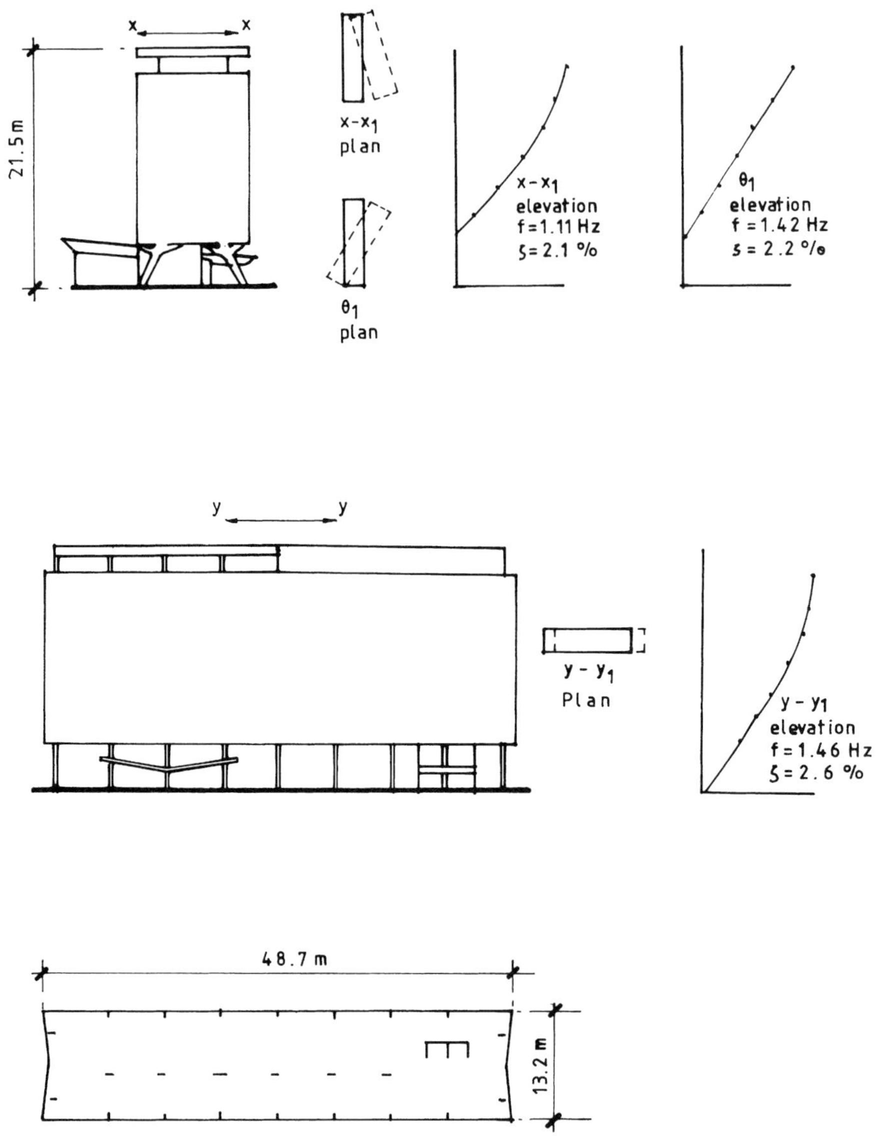

Fig. 6.11 Building in Oldham, UK.

An interesting test was conducted on the Engineering Tower at Leicester University in September 1979. As well as the induced vibration tests previously reported, use was made of the fact that the tower had a water tank to supply a Fluids Laboratory. The building was shaken both with the tank full and with it empty. The capacity of the tank was 17 000 gallons, which represents a mass of 0.077×10^6 kg.

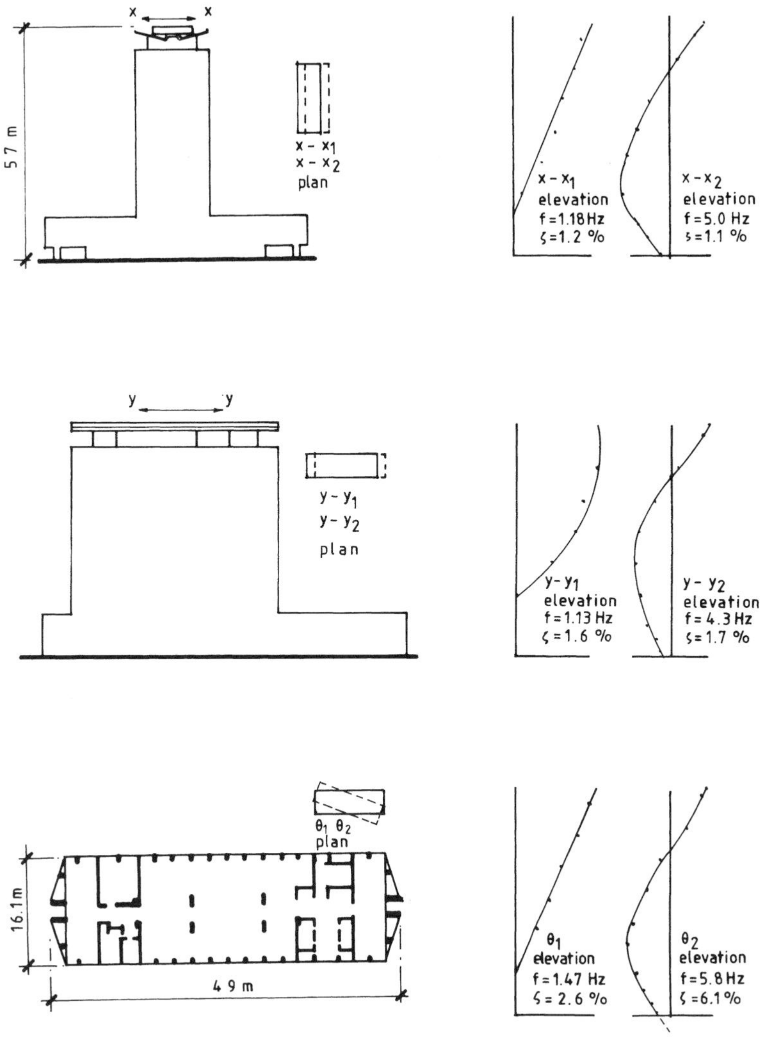

Fig. 6.12 Mode shapes for the Civic Centre, Plymouth, UK.

Whilst older buildings have taken a rectangular shape, more modern methods of slip-forming have allowed a more flexible approach to design. The building in Kuala Lumpur shown in Fig. 6.18 shows the use of diagonal bracing of concrete at one end of the building. This presents a lightweight alternative to the use of shear walls. Figures 6.19 and 6.20 show buildings constructed around the Kuala Lumpur city centre. The interesting point here is the variety of structural forms that are appearing. Each type of building form presents a new challenge to the dynamic structural designer in that the non-traditional appendages will all serve to move the dynamic

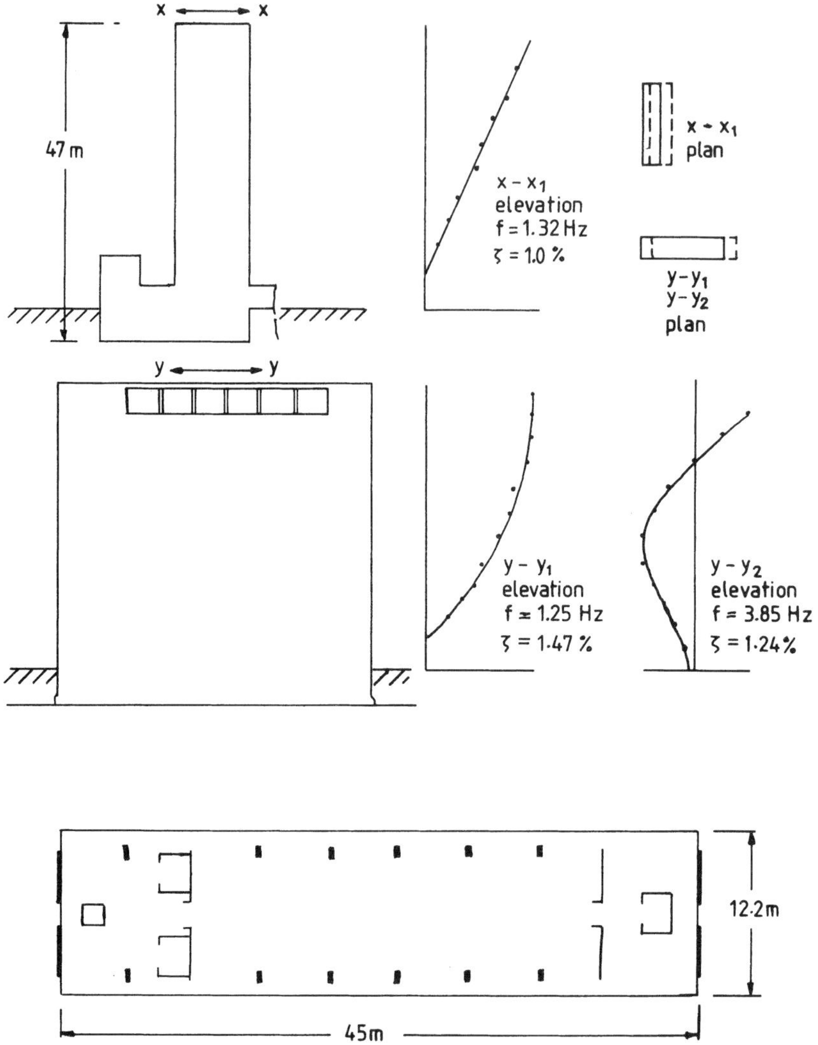

Fig. 6.13 Mode shapes of the British Rail building, Plymouth, UK.

characteristics a little further away from the norms associated with regularly shaped structures.

Perhaps the most interesting building in this area is the twin Petronas Towers of the world's tallest building. Figure 6.21 shows the building just before completion in 1995, and its form on completion in 1997 is shown in Fig. 6.22. The two towers are, despite the extravagant architecture, relatively straightforward in their dynamic response. However, the behaviour is

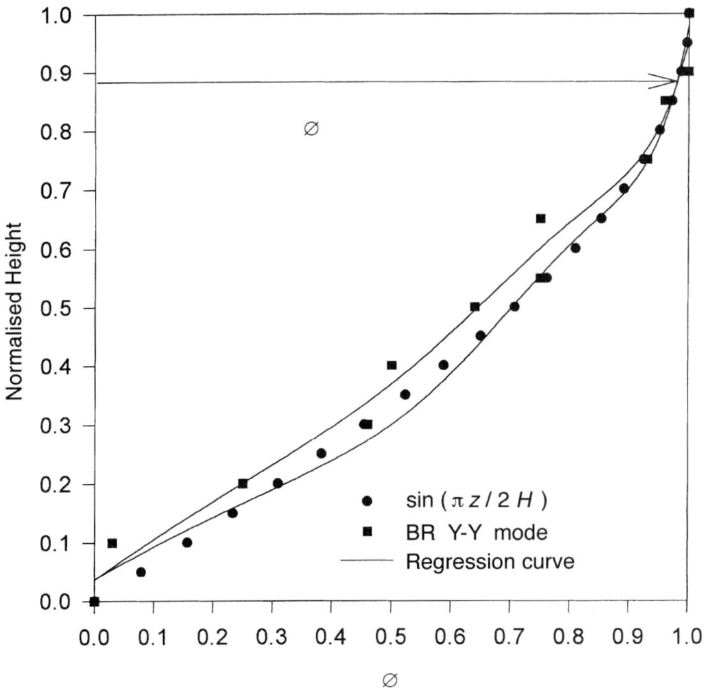

Fig. 6.14 Comparison of a measured and a theoretical shear-mode deflection, for the British Rail building, Plymouth.

complicated in an interesting way, by the provision of a skybridge at a height of 187 m above ground.

There are other interesting possibilities. Central Plaza in Hong Kong is basically a quasi-triangular plan form. Figure 6.23 shows public housing in Hong Kong, in which a non-traditional structural form has been chosen so as to maximize light penetration to the inner recesses of the structure.

Figure 6.24 is of a 'Clasp' system school building constructed in a coal-mining area in the UK. The large panels on the outer envelope are pinned at their top so as to allow for differential settlement. This particular form was tested under heavy vibration for the purpose of ascertaining the response in earthquake zones, and proved to be highly successful.

Variations to structural form have been changing rapidly. An early example is the Eiffel Tower in Paris. Figures 6.25 and 6.26 show two more interesting forms in Hong Kong, which have significant implications for the dynamic performance. Figure 6.27 shows the Bank of China Tower in Hong Kong. The position at which the central column stops is

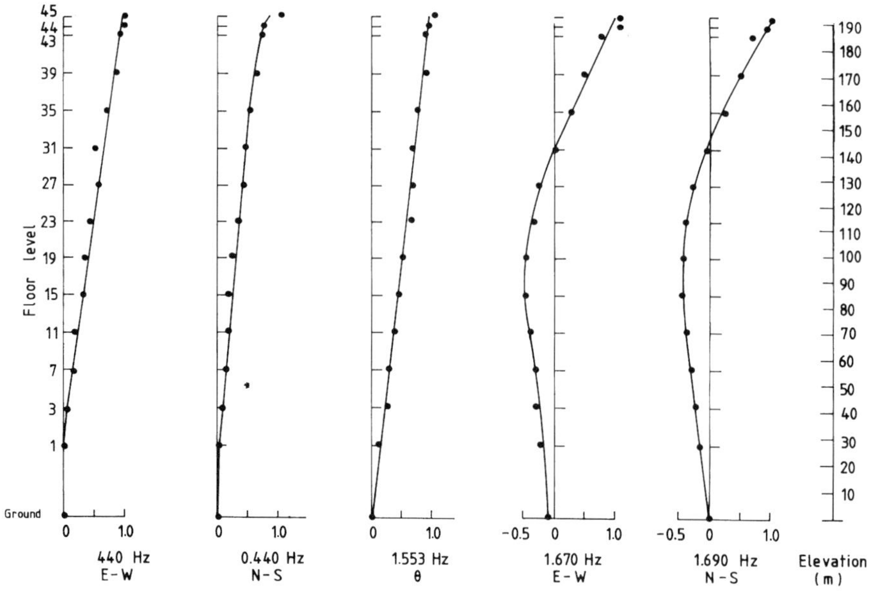

Fig. 6.15 Mode shapes of the NatWest Tower, London, UK.

shown in Fig. 6.28. It extends down from roof level to the 43rd floor, and there distributes load to the corner columns of the megastructure. The structure itself is an innovative combination of concrete and steel, in which the steel trusses are connected to each other only through the medium of the concrete. The inspiration for this unusual design concept may well have come from pagodas of the type shown in Fig. 6.29, in which a central column merely slides in a tube at ground level so as to avoid shear forces.

In the Citicorp building in New York City, a triangular form at the top of the structure contains an active vibration suppression device, for combating the effects of wind forces. In Fig. 6.30, the TransAmerica building in San Francisco is depicted. This particular structure is designed to withstand very large earthquakes.

6.2 *Chimneys*

In the case of chimneys, the torsional fundamental mode is rarely of significance because of the relatively small diameter of the chimney, and because the majority of the stiffness is concentrated around the periphery, making the structure torsionally strong.

The circular cross-section of chimneys would appear to make the prediction of modal directionality somewhat difficult. However, in practice, the directionality is dictated by appendages, and in particular flue ducts are

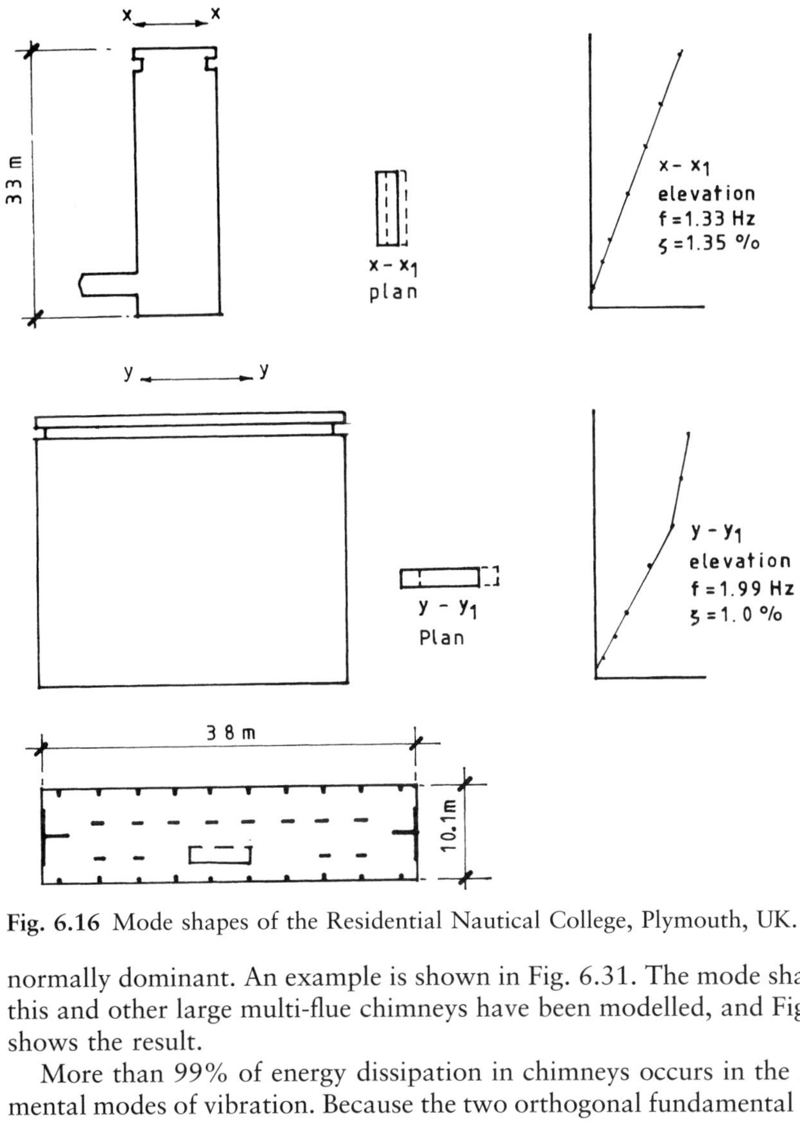

Fig. 6.16 Mode shapes of the Residential Nautical College, Plymouth, UK.

normally dominant. An example is shown in Fig. 6.31. The mode shapes of this and other large multi-flue chimneys have been modelled, and Fig. 6.32 shows the result.

More than 99% of energy dissipation in chimneys occurs in the fundamental modes of vibration. Because the two orthogonal fundamental modes of vibration are inevitably very close in frequency (normally only thousandths of a hertz difference), then energy is swapped very easily between them. Figure 6.33 gives an example taken from measurements of the tip of a nearly 200 m tall chimney.

6.3 Bridges

The dynamic response of bridges is quite complex. They are truly three-dimensional structures and their dynamic behaviour reflects this fact. As a

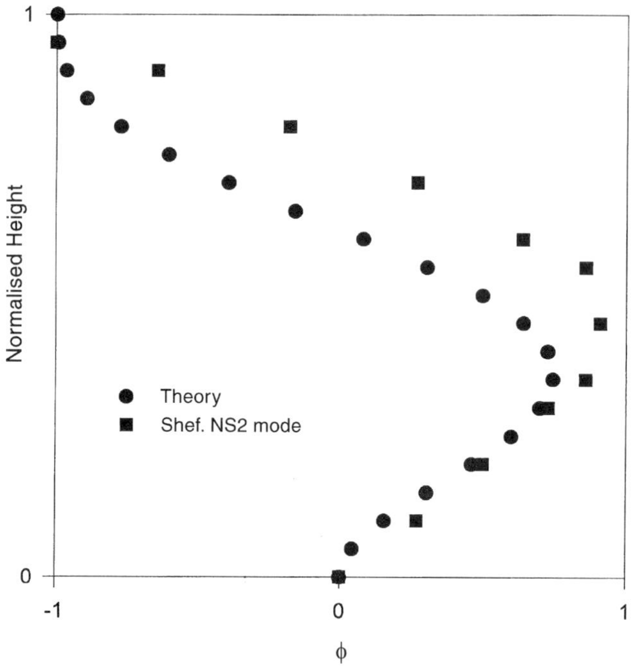

Fig. 6.17 A comparison of a measured and a theoretical curve for a second translational mode (tall building) and NS2 mode, for the Sheffield University Arts Tower.

result, there are three directions in which the dynamic response must be described:

- Vertical (V)
- Upstream–downstream (UD)
- Across-valley (AV)

Across-valley modes tend to be much stiffer than in the other two directions, because they must interact with valley (or embankment) walls or involve material 'breathing-type' modes. Such a mode of vibration can be envisaged by breathing deeply and observing the behaviour of your own chest.

In both the vertical and the upstream–downstream directions, the mode shapes take on a similar aspect when committed to a two-dimensional representation. Many of these modes of vibration will also occur at similar frequencies, and so the ability of a bridge to couple the response in these two directions is quite large. The behaviour of the Tacoma Narrows Bridge just prior to its collapse showed clear signs of a coupling of vibration from the V and UD directions.

It is worth noting that in the V and UD directions the lowest-frequency mode involves a 'breathing-type' action, and the result is that such modes

Fig. 6.18 Inventiveness with concrete, Kuala Lumpur.

do not normally contribute significantly to the response. Having observed this, however, it must be pointed out that as many as 10 modes of vibration, in both V and UD directions, must be included if a realistic estimate of the response of a bridge is to be achieved. Once again, modal coupling must be accounted for in the response of bridges, and the approach taken for tall buildings is recommended.

Figure 6.34 shows the Clifton Suspension Bridge. This bridge is noteworthy in that, whilst it was completed in 1864, the start of the project was in 1837. At the earliest stages, there was much debate about the span length, and whether the bridge could in fact withstand the wind load. Various designs tried to address this problem. Eventually, after much discussion, Brunel's design won the day, and has proved its ability to withstand the wind action by performing satisfactorily for the intervening 130 years. A later development was to invert the suspension mechanism and use the arch action. This allows the use of concrete, which is ideal in this situation. Figure 6.35 shows the example of the William Jolley Bridge in Brisbane, Australia.

Fig. 6.19 Kuala Lumpur city centre, showing the variety of structural forms.

Fig. 6.20 Kuala Lumpur city centre.

Fig. 6.21 Petronas Towers during construction, 1995.

Fig. 6.22 Petronas Towers, 1997.

Fig. 6.23 Domestic housing, Hong Kong, non-traditional structural form.

Fig. 6.24 A 'Clasp' system school, UK.

Fig. 6.25 The Lippo Centre, Hong Kong, with construction of the Bank of China Tower.

In practice, the span length of bridges is the dominant parameter in determining the fundamental natural frequency of bridges. As the span length increases to greater than 30 m, then the natural frequency becomes predictable with a high degree of certainty from simple formulae. Figure 6.36 summarizes the results of experiments conducted in Belgium, Switzerland and the UK. The formula suggested for the prediction of the fundamental natural frequency is

$$f_0 = L^{-0.9} \tag{6.1}$$

where L is the span length. This formula gives good predictions for the longest spans in the world, such as the Golden Gate Bridge, San Francisco, and the Tsing Ma Bridge, Hong Kong.

For more common types of bridge, such as that in Maranhao, Brazil (Fig. 6.37), then the action of traffic or trains is of fundamental importance. Figure 6.38 is a computer-based simulation of the response of the Maranhao

Fig. 6.26 Hongkong and Shanghai Bank building, Hong Kong.

bridge to the passing of an ore-laden train at a speed of 25 m/s. In cases such as this, the velocity of the train effectively dictates the excitation frequency of the bridge. The quasi-static deflection of the bridge is also noteworthy. The deflected vertical mode form for this case is shown in Fig. 6.39.

In earthquake-prone areas, anti-seismic devices are normally added for the protection of bridges. The rail bridge in Atsugi City, Japan (Fig. 6.40), is such an example, and the mounting detail is shown in Fig. 6.41.

6.4 Dams

In Chapter 4, the theoretical shapes for modes of vibration for dams were discussed. In practice, the real behaviour of dams is a little more complex, and is dominated by the water in the reservoir and particularly its level relative to the dam. Figure 6.42 shows the reservoir behind the Contra Dam in Switzerland (Fig. 6.43), at the start of a series of induced vibration tests. This testing was suspended after a severe rainstorm and lightning strikes on the dam. This resulted in the raising of the reservoir water level by 20 m in

Fig. 6.27 Bank of China Tower, Hong Kong.

a period of 12 h, and a rise to 2 m above spillway level (Figs 6.44 and 6.45). The enormous power of such an event raises vibration levels above the perception level at times.

It is normally assumed that the dam will behave monolithically, and when the reservoir level is high, then this is the case. However, dams are not constructed in a manner that encourages monolithic behaviour. Figure 6.46 shows the Hoover Dam in the USA, and it is possible to discern where the slipforms for the pouring of concrete have been placed. The reason for constructing in this manner is that, with the large amounts of bulk concrete used, then sufficient opportunity for dissipation of heat has to be given, and this dictates that the dam will be constructed in sections. Movement between these sections then becomes probable.

Fig. 6.28 The central column of the Bank of China Tower.

Fig. 6.29 A pagoda, Hong Kong.

Fig. 6.30 The TransAmerica building, San Francisco.

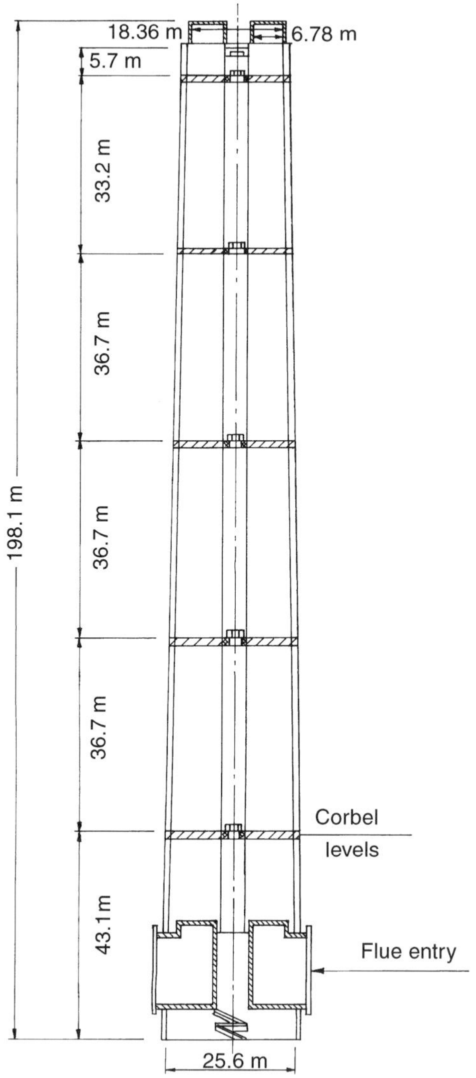

Fig. 6.31 Section of a multi-flue chimney.

In the case of gravity dams or buttress dams, individual sections of the dam may have seals between them, in recognition of the fact that relative movement will take place. Figure 6.47 shows an example of a buttress-supported dam.

In all such cases, the raising of the reservoir water level acts to make the dam behave more as a monolith. The reason for this behaviour is probably

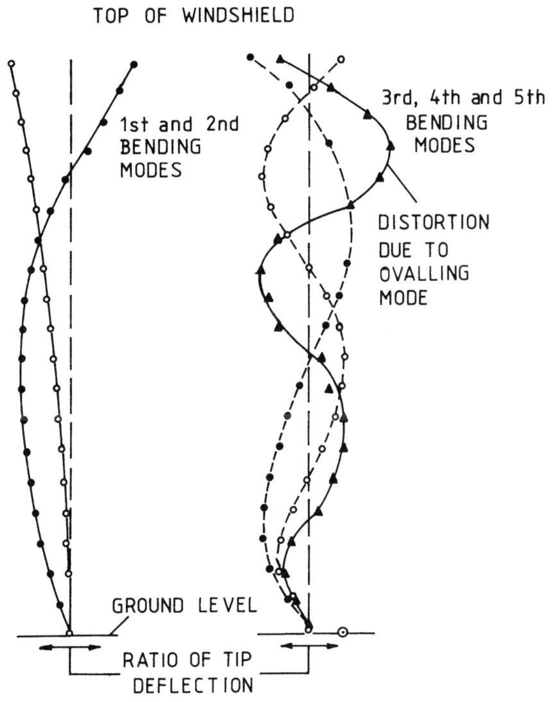

Fig. 6.32 Mode shapes measured from a model of a multi-flue chimney.

that even a small amount of twisting of a section will tend to make adjacent units lock together.

The case of Wimbleball Dam is particularly interesting from this point of view. The dam is a concrete buttress type, and is shown in Fig. 6.47. When the reservoir level is lower, then these individual sections behave independently. Figure 6.48 shows the response of the dam, which was recorded as a series of synchronized vibrators, set on the crest of the dam to induce motion in the fundamental upstream–downstream mode of vibration, were switched off. It can be seen that each buttress undergoes an individual response that is different from its neighbours. It can also be seen that in some cases the response becomes greater before decaying. Such an increase can only occur if energy is transferred in from one of the other buttresses.

In the case of the Wimbleball Dam, the fundamental natural frequencies changed with the raising of the reservoir water level. This is shown in Fig. 6.49. Such a change of natural frequency with the water height was also noted for the Contra Dam in Switzerland (Fig. 6.43), when the storm referred to previously raised the water level of the reservoir from a height of 454.05 m to 470.97 m. The fundamental natural frequency was tracked

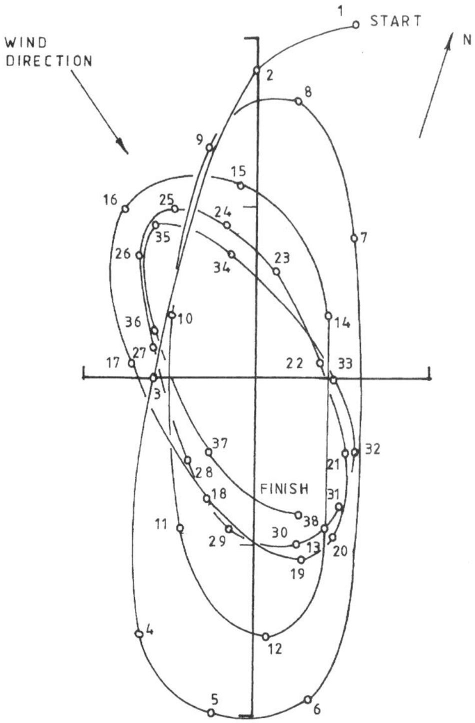

Fig. 6.33 The motion of the top of a chimney over a short period.

Fig. 6.34 The Clifton Suspension Bridge, Bristol, UK.

Fig. 6.35 Concrete bridge in Brisbane, Australia.

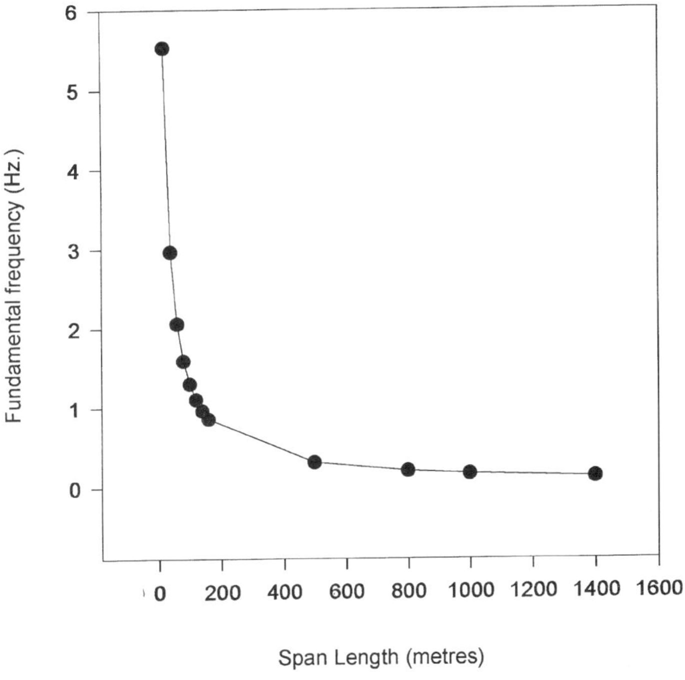

Fig. 6.36 Fundamental natural frequencies of bridges.

Fig. 6.37 Railway bridge, Maranhao, Brazil.

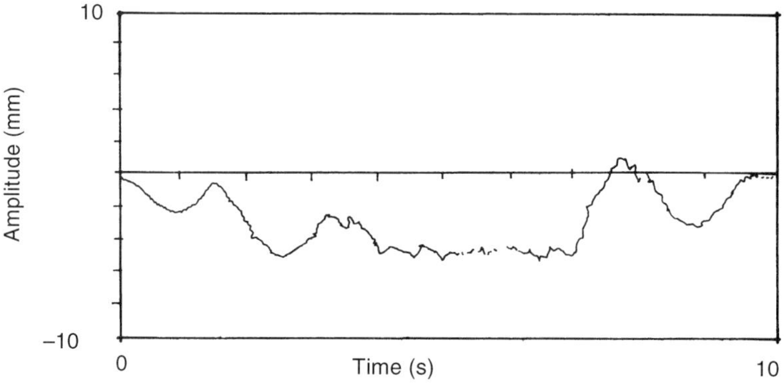

Fig. 6.38 Response of a bridge to a traverse of a train at 25 m/s.

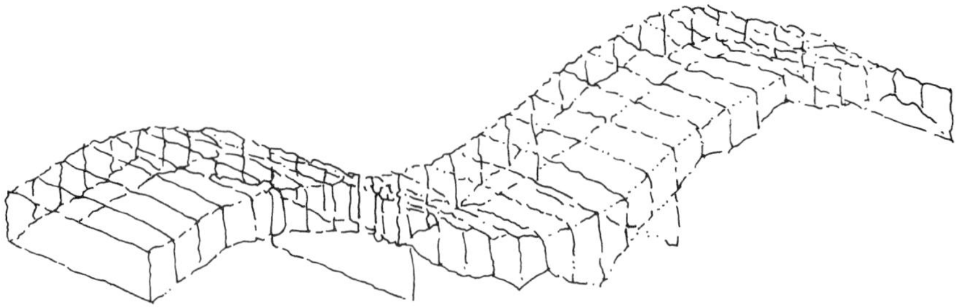

Fig. 6.39 Deflected mode shape of a railway bridge, Maranhao, Brazil.

Fig. 6.40 Railway bridge, Atsugi City, Japan.

Fig. 6.41 Bearing detail of the Atsugi City railway bridge.

Fig. 6.42 Contra Reservoir.

Fig. 6.43 Contra Dam, Ticino, Switzerland.

throughout this rise, and the result is shown in Fig. 6.50 as a natural frequency–water height relationship. In this particular case, such a measurement was possible because of a very rapid rise in the water level after a rainstorm and a particularly high rate of run-off of water around the reservoir valley.

Fig. 6.44 Overspill at Contra Dam.

Fig. 6.45 Overspill viewed along the crest at Contra Dam.

Dams themselves form part of the scenery in a valley. That is to say, the boundary of the dam system is not as well defined as in the case of a bridge or a tall building. It is clear from tests conducted on dams using a vibrator system that the dam itself induces motion in the surrounding countryside.

Fig. 6.46 Hoover Dam, USA.

Fig. 6.47 Wimbleball Dam, UK.

Such movements are measurable, and Fig. 6.51 shows an example of such measurements at Brianne Dam in Wales. In this particular case, the dam was being shaken in a fundamental mode, and the vibrations continued several tens of metres past the supposed edge of the dam. In the case that

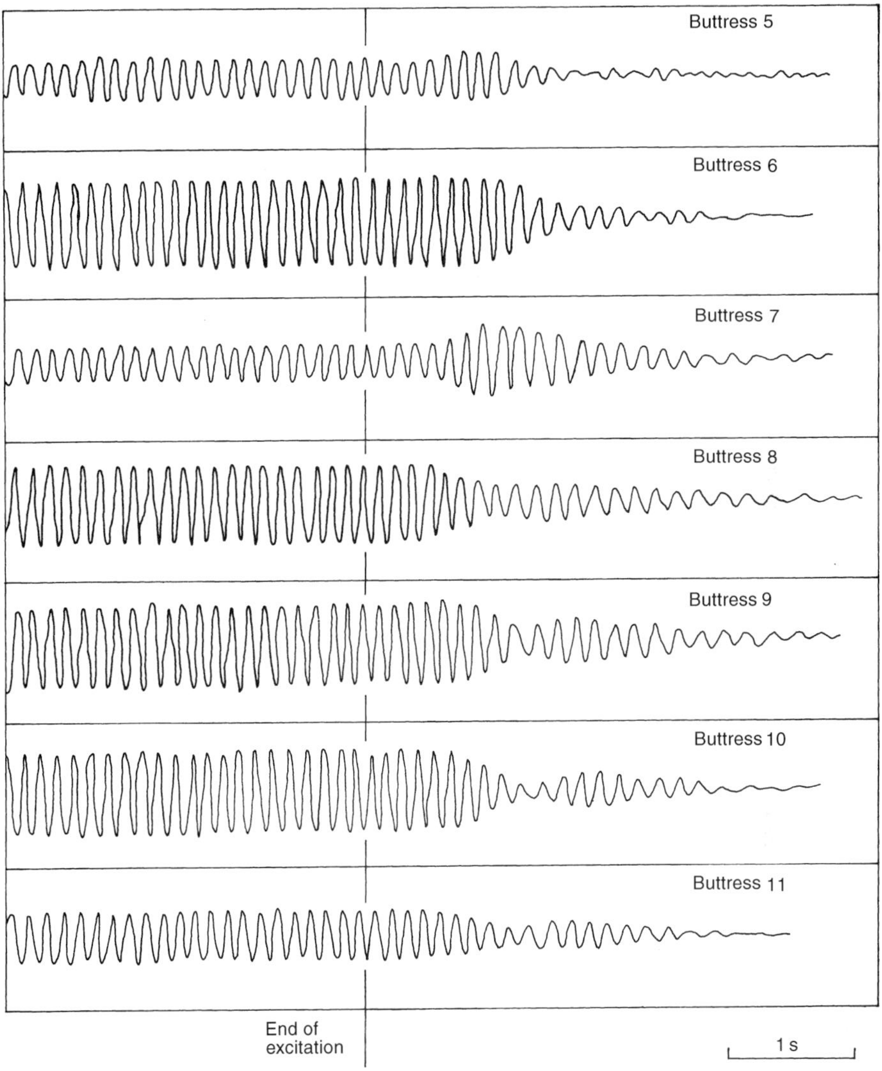

Fig. 6.48 Response decay on buttresses 5–11 after shutdown of excitation of buttress 8, at Wimbleball Dam.

the vibration source is off the dam, then clearly motion in the surrounding countryside will be transferred into the dam. This observation has implications for the calculation of forces induced by earthquakes, since the energy appearing in the dam will be collected from a wider area than just the 'footprint' of the dam. This factor emphasizes the importance of using the modal influence function when calculating the action on the dam.

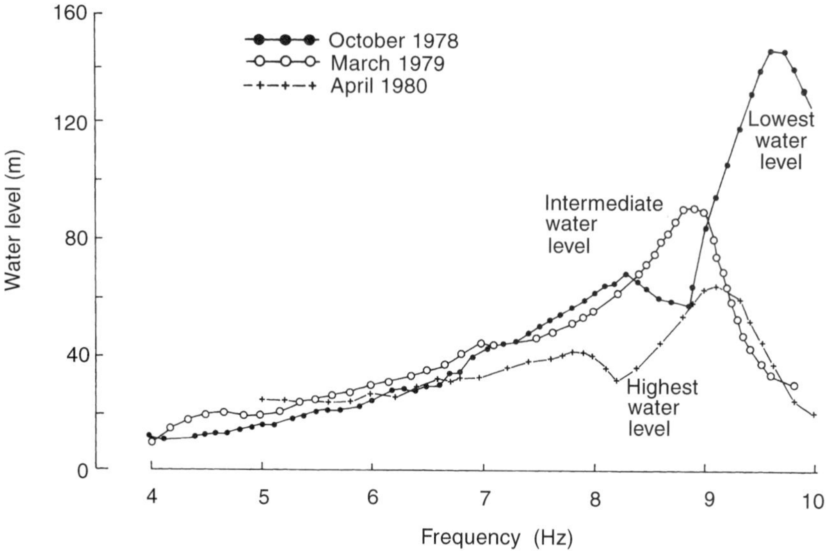

Fig. 6.49 Spectra taken from Wimbleball Dam.

One of the important parameters necessary for the calculation of the response of dams is the effect of the added mass caused by the existence of the water in the reservoir. The calculation of this added mass for the Emosson Dam is included as Appendix G. In this particular case, at the time of testing, the water level was 7 m below the level of the spillway. The mass of the dam measured from the induced vibration test was 20.55×10^8 kg, and that inferred from the design drawings is 22.18×10^8 kg. It is often assumed that a mass equivalent to 40% of the mass of the dam needs to be added to account for the mass of water being moved by the motion of the dam. Clearly, in this case, that theory is not confirmed. However, it must be remembered that the water level was 7 m below spillway, and that the effect may be more pronounced when the water level is at its maximum. No further tests on dams under induced vibration have been conducted in the last 20 years, and so it is still not possible to confirm whether the 40% added mass theory is correct or not.

The characteristics of dams are independent, in general, of the materials of which they are composed. Earthfill or rockfill dams such as Brianne (Fig. 6.52) and Plover Cove, Hong Kong, behave as monoliths. The mode shapes measured at Brianne Dan are presented in Fig. 6.51.

Very large arch dams like Contra behave in a well defined way, especially when the water level is high. Similar examples come from Emosson (Fig. 6.53) and Zervreila, both in Switzerland. A comparison of the modes of

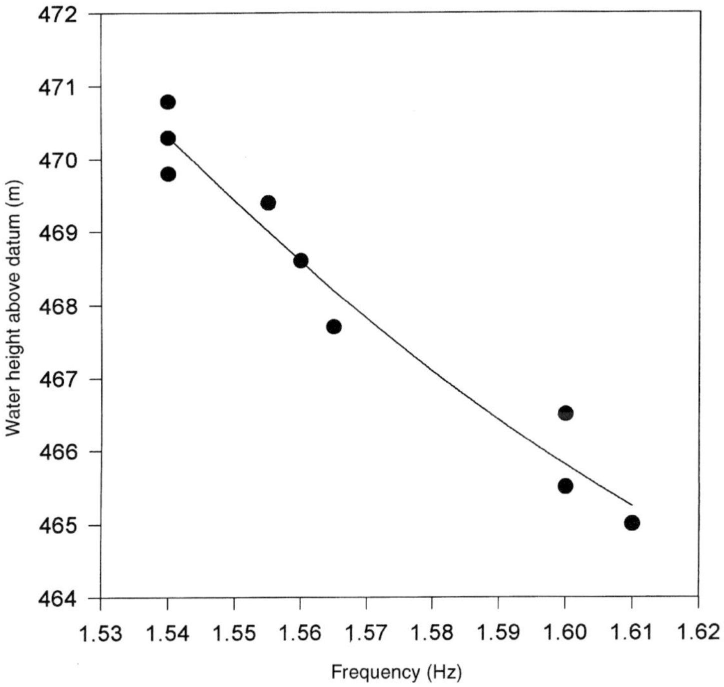

Fig. 6.50 Frequency versus reservoir water height at Contra Dam, Ticino, Switzerland.

vibration excited in a forced vibration test at Emosson was made with a three-dimensional finite-element simulation, and the results were very similar. The results of the mode shape and frequency comparisons are presented in Figs 6.54–6.56.

In general terms, for all types of dam, there is a relationship between frequency and the height of the dam itself. This relationship is presented in Fig. 6.57, and serves as a guide to the choice of these parameters for design. It is also possible to produce a similar result for the value of the fundamental mode damping value for dams, and this is presented in Fig. 6.58. As a result, it is possible to produce a relationship between damping and frequency for dams in general. This is shown in Fig. 6.59. This result is quite remarkable, because it appears to apply to all structures in general. Not only is this general form found for dams, but it also applies to tall buildings, chimneys and bridges. Stated in very simple terms, the low-amplitude value of damping for the fundamental mode (expressed as a percentage) is numerically equal to its natural frequency.

Lastly, there is a further source of excitation of dams – that of wave action. Although the action itself produces relatively small forces, the

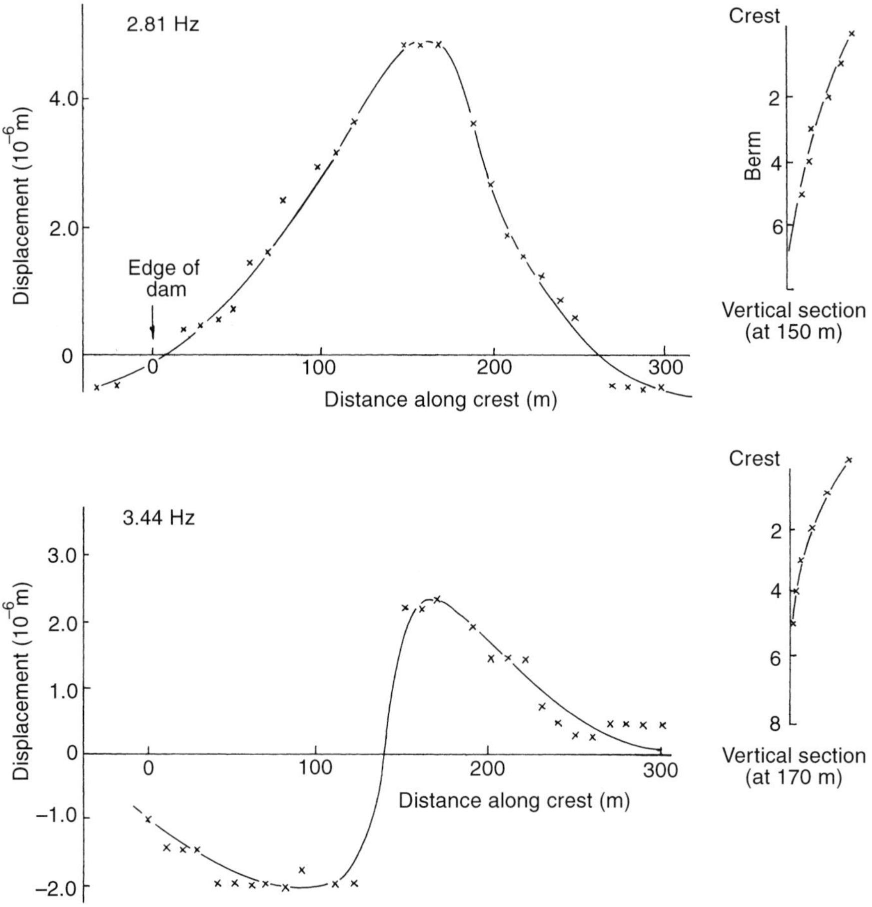

Fig. 6.51 Mode shapes measured at crest of Brianne Dam.

presence or absence of such an effect makes for an intriguing difference to the visual aspects of a dam. Figure 6.60 shows the Upper Glendevon Dam under a wind-induced down-valley wave action.

6.5 Offshore structures

Offshore structures are often similar in general form to a tall building or a chimney. However, the similarity in response ends at that point. The mass of water around the structure has to be moved with it, and so the dynamic system includes an added mass of water. Additionally, there is a zone in which the action of waves is important. Large voids in the structure are occasionally filled with oil or with gas. This in turn affects the mass of the

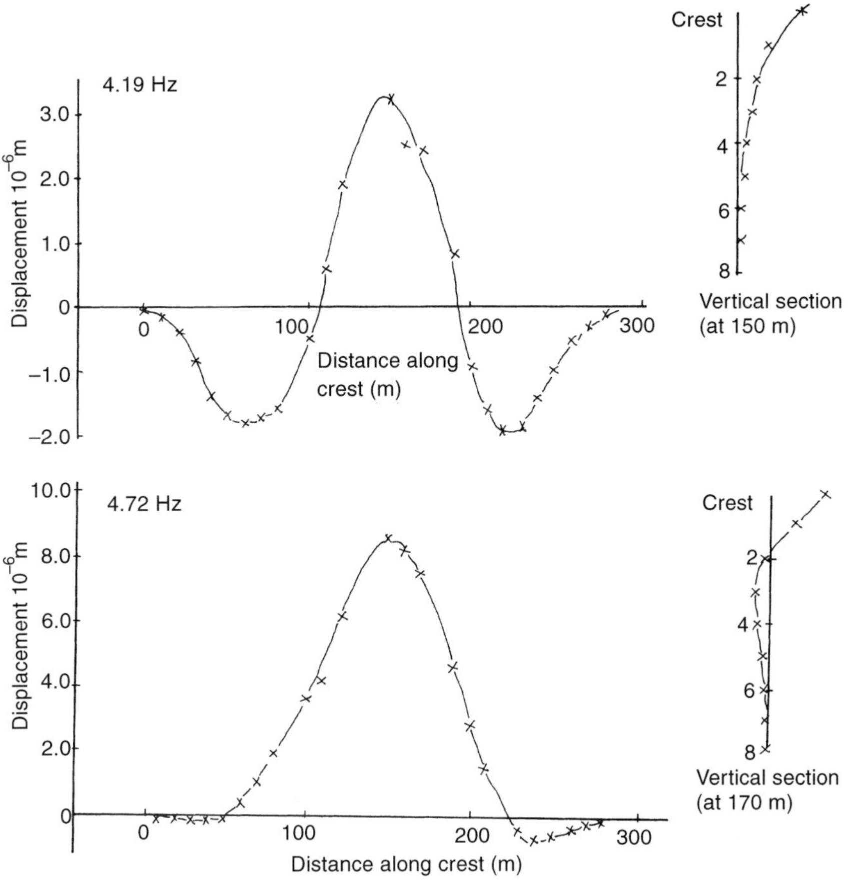

Fig. 6.51 *Cont'd.*

structure. However, since these locations are normally near to the base, then the effect on the modal mass is relatively small.

In many cases, local modes of vibration of parts of offshore structures become important. Appendages such as access platforms or individual structural members may have resonance frequencies that are low enough to be comparable with the natural frequencies of the entire structure. This factor makes the use of dynamics for the interpretation of the health of a structure a doubtful proposition, since the overall behaviour can be confused with that of a smaller element.

In the 1970s, as part of an initiative to try to ascertain the loading on offshore structures, an experimental model was deployed for the purpose of making measurements. Figure 6.61 shows the original rig being prepared in a dry dock. Shortly after deployment, the structure was struck by a storm

Fig. 6.52 Brianne Dam, Wales, UK.

(Fig. 6.62) and was damaged. The structure was repaired and redeployed as shown in Fig. 6.63. The deflected mode shapes are shown in Fig. 6.64, and the measurement of the spectrum of the response is shown in Fig. 6.65. This spectrum is particularly interesting, because the structural responses can be seen in the region of 4–6 Hz, whilst the enhanced response below 1 Hz is the forced spectrum of wave loading caused by the action of the sea.

6.6 *Floors*

It is often considered that the principal function of floors is merely to withstand the dead load imposed on them. Whilst this is an important aspect of the behaviour of a floor, a further requirement is that they must withstand dynamic loading. This dynamic loading can be imposed by people or machinery. In all cases in which a dynamic problem occurs, the

Fig. 6.53 Emosson Dam, Switzerland.

response is generated by the forcing frequency coinciding with the resonance frequency of the floor.

The form of excitation caused by a person jumping is shown in Fig. 6.66. This waveform was produced by the author jumping up and down on a force transducer. The force depicted here is representative of 350 people dancing in unison. The frequency at which this occurs is a function of the speed of jumping. The most natural frequency (which corresponds with the speed of walking) is in the region of 1–2 cycles per second. When a person is jumping, then there is an added impulse that is caused by the impact of the heel on the floor. Because of the non-sinusoidal form of the input forcing function, then harmonics exist. This implies that, at double and four times the fundamental forcing frequency, there is a smaller forcing function that may excite a frequency of the floor. This factor is important for smaller floors in which harmonics may excite the resonance of higher frequencies.

The forcing function in Fig. 6.66 was applied to a floor having a natural frequency of 5.79 Hz and 4.4% damping. The response is also shown in Fig. 6.66. In this, the forced response at the frequency of jumping is clearly seen, whilst the more frequent and smaller vibrations at the resonance frequency are also seen. This example comes from tests conducted at Middlesbrough Town Hall, UK. Figure 6.67 shows the stage area and the floor. The natural frequency and damping of the floor were measured by shaking with a small vibrator, and this response is shown in Fig. 6.68. This established the parameters to be used. The Duhamel integral was then used, with the forcing function $F(t)$ represented by a time history of the measured

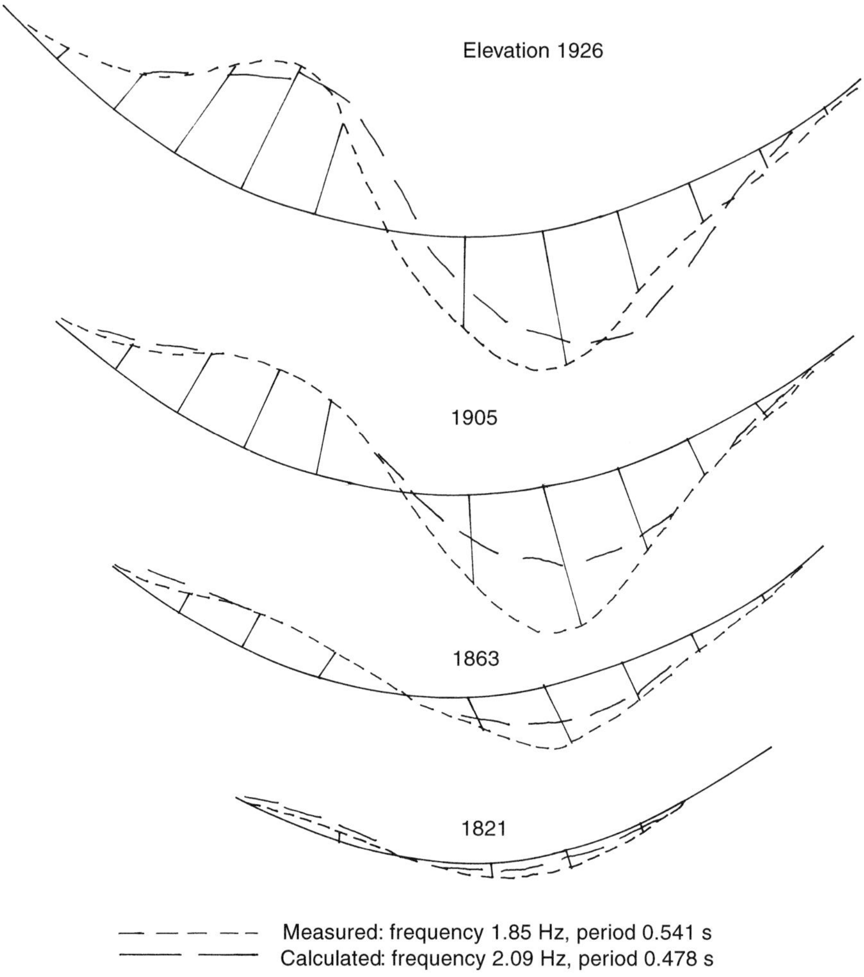

Fig. 6.54 Comparison of measured and predicted fundamental mode at Emosson Dam.

forcing function proportioned to represent 350 people. This was modified to become a modal forcing function by using the modal influence function. The frequency of the input was also varied to simulate different circumstances. Figure 6.69 shows the result of these calculations at one-quarter resonance frequency (1.928 Hz) and one-half resonance frequency (2.968 Hz). (The resonance frequency varies with the amplitude.) The response at one-half resonance frequency is seen to be very much enhanced as the crowd impacts with the floor every half-cycle. The response built up to a large maximum because of the short time constant.

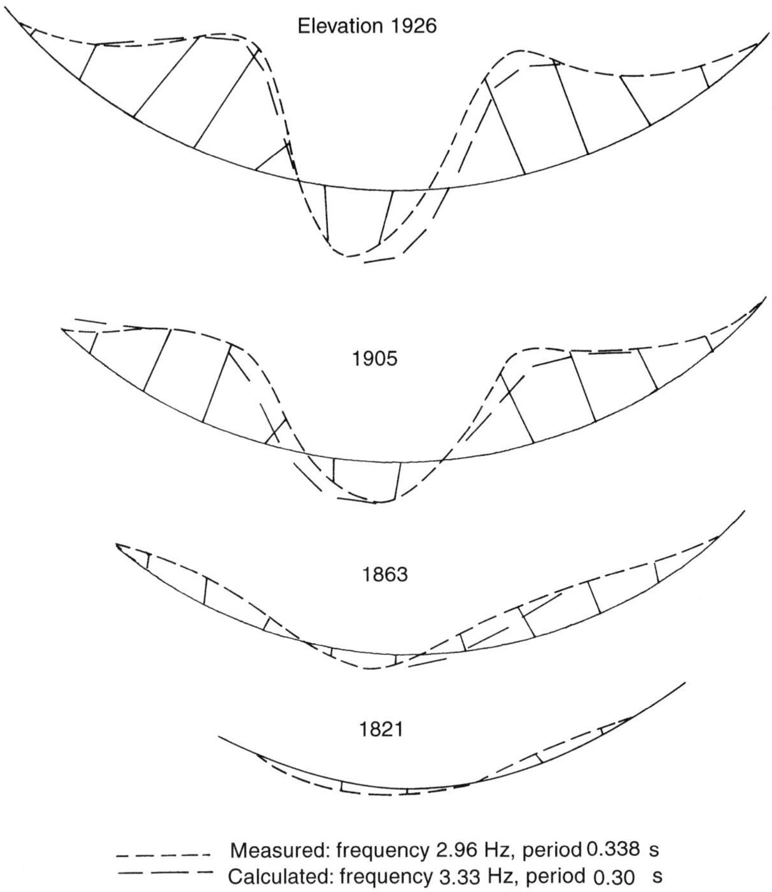

Fig. 6.55 Comparison of measured and predicted second mode at Emosson Dam.

Mode shapes for the Middlesbrough floor, both with an added mass (intended to represent the mass of a crowd) and without the added mass, are shown in Figs 6.70 and 6.71. These figures can also be used as influence functions.

When larger crowds are on floors and they jump up and down in synchronization, then the entire dynamic response becomes more complex. Firstly, the mass of the floor will change throughout each half-cycle of the response of the floor. The frequency will be a function of the mass of the floor and half of the mass of the crowd. The people jumping will also increase the ability of the floor to dissipate energy, and accordingly damping will rise. However, the initial level of damping of a floor is fundamentally dependent on the materials used and the boundary conditions. In

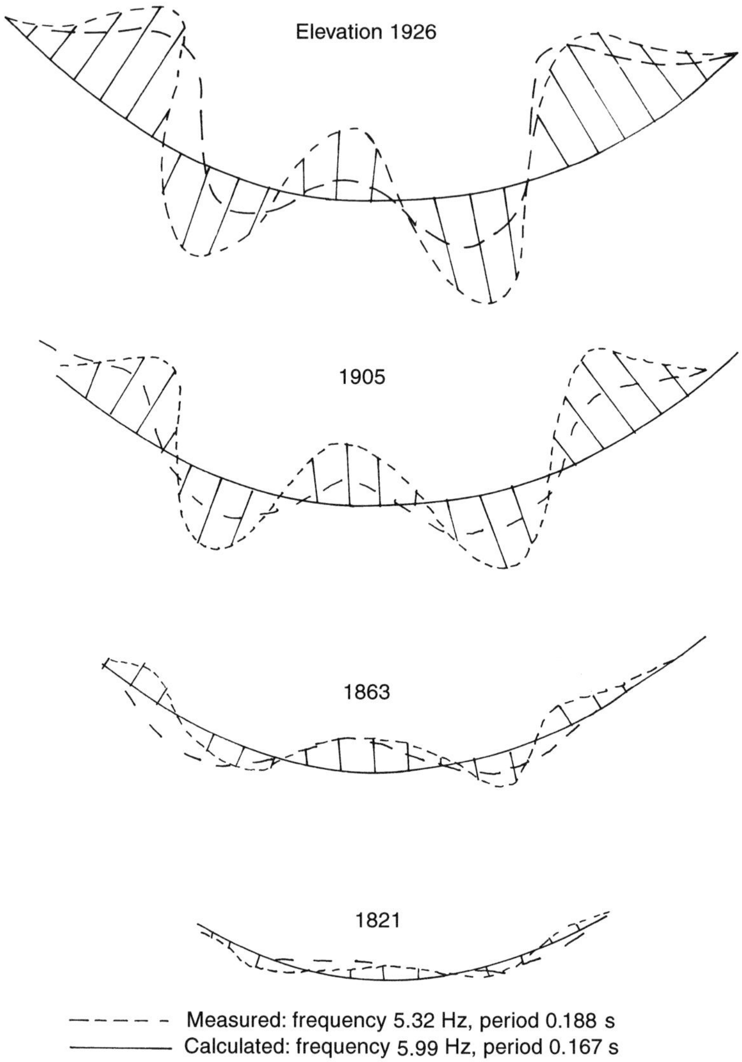

Fig. 6.56 Comparison of measured and predicted mode shapes for a complex mode at Emosson Dam.

the example above, none of these reductions were considered, because it was intended that the calculated response should err on the conservative side. Subsequent measurement of the response of the floor to a real crowd under concert conditions showed that the conservative assumptions were justified. The implication of this observation is that the synchronization of crowds tends to be very effective, and that energy taken by the crowd is very largely returned to the floor.

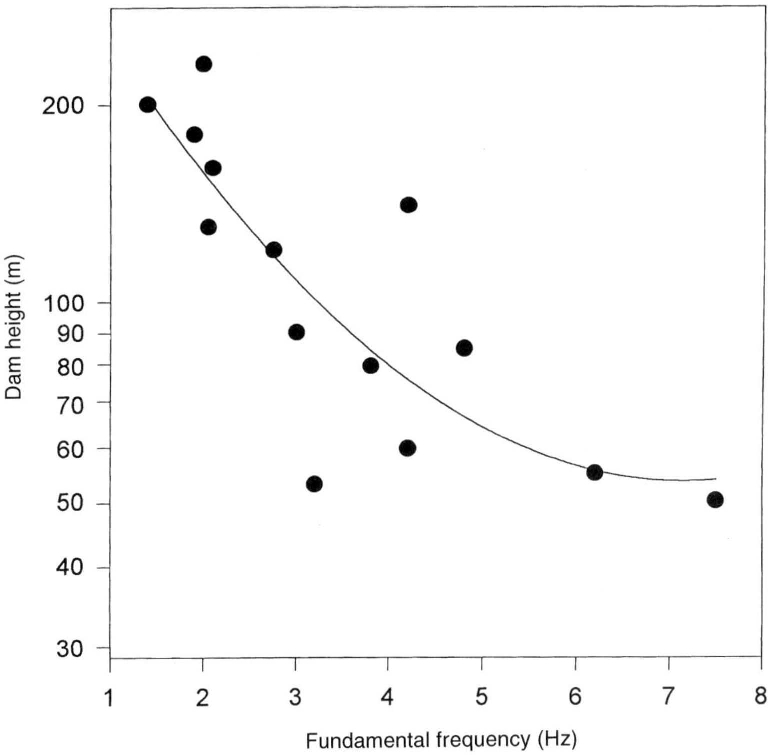

Fig. 6.57 Height versus frequency characteristic for dams.

In general terms, the acceptable limits to deflection of a floor can be considered to be span/250.

In the case of machine-induced vibration, the situation is more easily defined. In this case, there are two types of machine-induced vibration. The former is caused when there are unbalanced rotating weights in the machinery, whilst the latter is caused when the machine induces a percussive effect through a part of the machine striking another surface.

In the case of reciprocating machinery, the forcing caused by the unbalanced weight is simply

$$mr\omega^2$$

where m is the unbalanced mass, r is the eccentricity of the mass and ω is the rate of rotation. Of course it is not always possible to know what the magnitude of the unbalanced mass is. In these circumstances, measurement of the response is necessary.

In the case of percussive machine-induced vibration, measurement is recommended. In this case, the measurement of vibration on a floor will normally show a response that is an inverse square function of the distance

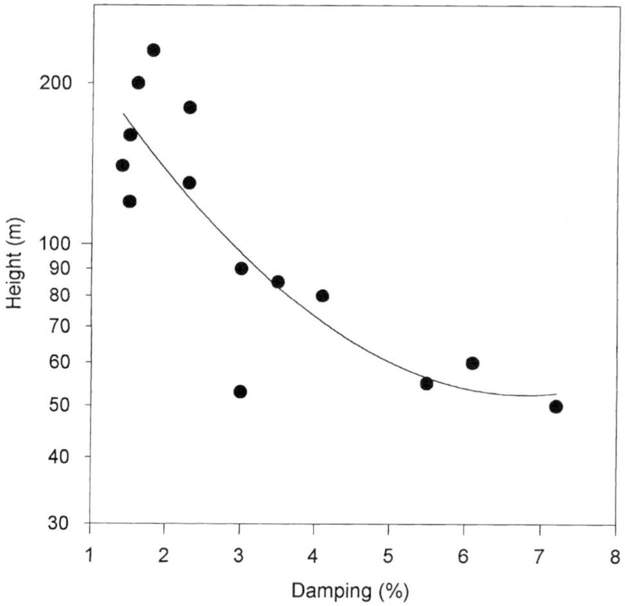

Fig. 6.58 Height of dam against damping.

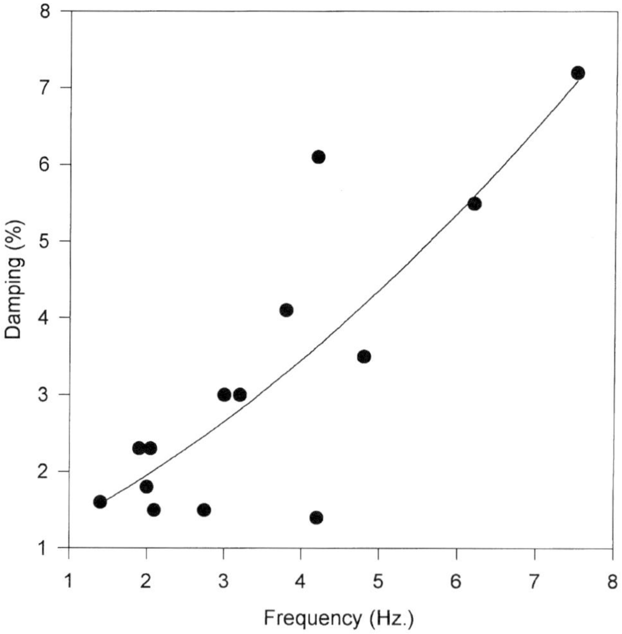

Fig. 6.59 Damping versus frequency for all dams.

Fig. 6.60 Upper Glendevon Dam, subject to wind and wave action.

Fig. 6.61 OSFLAG experimental offshore structure at Blackwell Dock.

Fig. 6.62 OSFLAG experimental structure during a storm. (Source: Building Research Establishment.)

from the source. If, however, the response of the floor is at a resonance frequency, then the floor integrates the energy input and responds with a normal mode of vibration in response.

6.7 Coupling of modes of vibration

Figure 6.72 shows a measurement of the positioning of a fundamental mode and its orthogonal pair for a tall building. What is striking about this diagram is that the modal pair does not line up with the geometry of the building. In practice, the lowest-frequency fundamental mode seeks out the direction of least stiffness of the building. Normally, this will align with the geometry, but occasionally it will not when there is a severe axisymmetry of the stiffness, as is the case with the building in Fig. 6.72. Having established the fundamental direction, the orthogonal fundamental mode must be at 90° to that direction. It is worth noting that, in the case of irregularly shaped buildings, it is extremely difficult to predict the directionality of the fundamental mode direction.

The two fundamental modes of vibration are joined by a torsional fundamental mode. It just so happens that, for the case of tall buildings of conventional shape, size and stiffness, the three fundamental modes of vibration are often very close in frequency. A frequency spectrum for the Sheffield Arts Tower was shown in Fig. 6.1, in which three spectral peaks representing these three fundamental modes of vibration are clearly visible. In this particular case, the three modes of vibration all occur within a

Fig. 6.63 OSFLAG replacement structure. (Source: Building Research Establishment.)

frequency interval spanning only one-quarter of a cycle per second. With such a close frequency proximity, it is very easy for a structure to switch energy from one mode of vibration to another, through geometric coupling.

6.8 Gas explosions

Finally, an example of a gas explosion and its consequences is included here. The main purpose of this example is to emphasize that design and codification is inexact, and has many traps for the unwary.

Figure 6.73 shows the Ronan Point building shortly after a gas explosion on the eighteenth floor of the 22-storey block caused a progressive collapse. The building was repaired with an extra large column in the corner that collapsed. Unfortunately, some 15 years later, another problem with the building suggested that it was 'coming apart at the seams'. Tests were

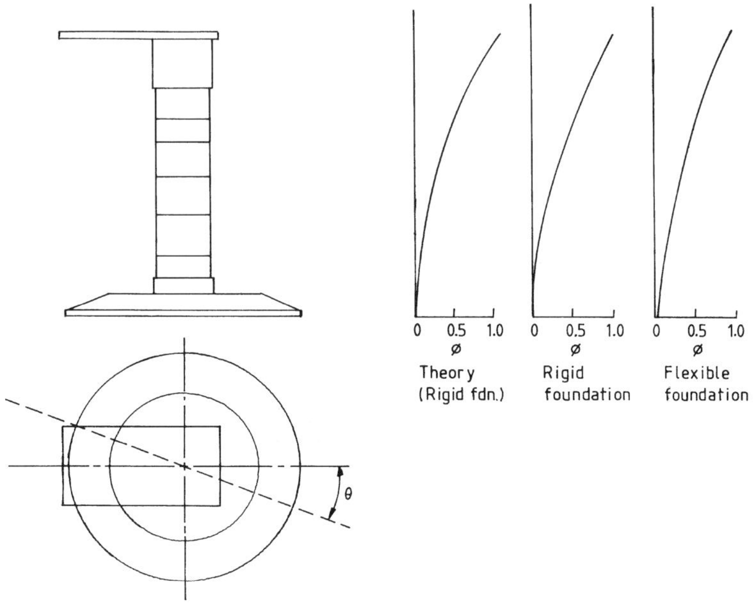

Fig. 6.64 Normalized modes for a model of the OSFLAG offshore structure.

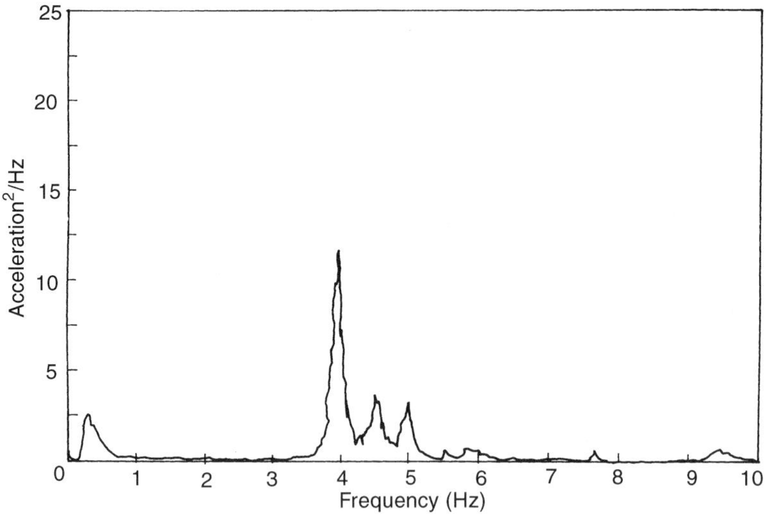

Fig. 6.65 Spectrum from the OSFLAG structure at sea: the power spectral density for a particular direction (105°).

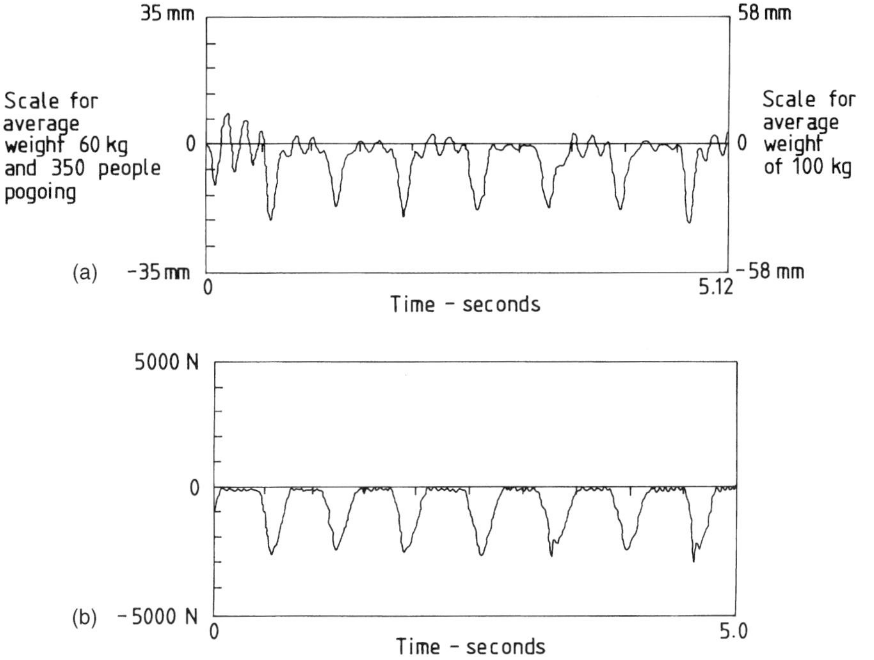

Fig. 6.66 (a) Pogo dancing forcing function and (b) the response of a floor system. The system has a frequency of 5.79 Hz and 4.4% damping.

Fig. 6.67 Stage and dance floor at Middlesbrough Town Hall.

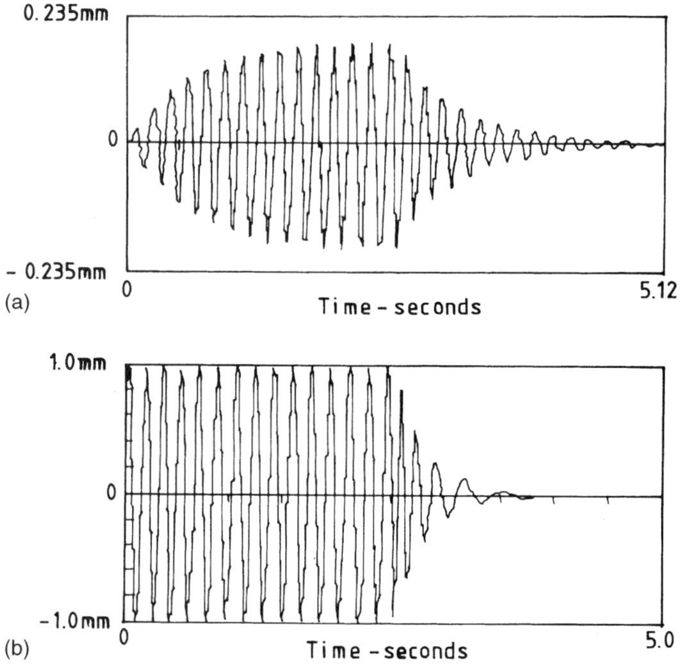

Fig. 6.68 (a) Vibrator-excited forcing and (b) response of the Middlesbrough floor. The system has a frequency of 5.79 Hz and 4.4% damping.

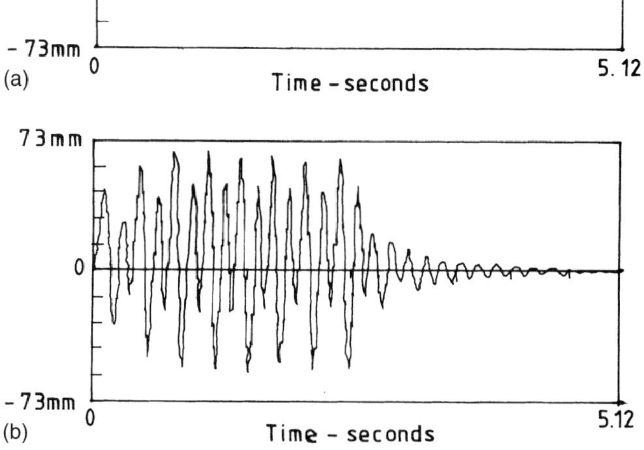

Fig. 6.69 Response of the Middlesbrough floor to forcing at different frequencies: approximately (a) one-quarter and (b) one-half resonance frequency.

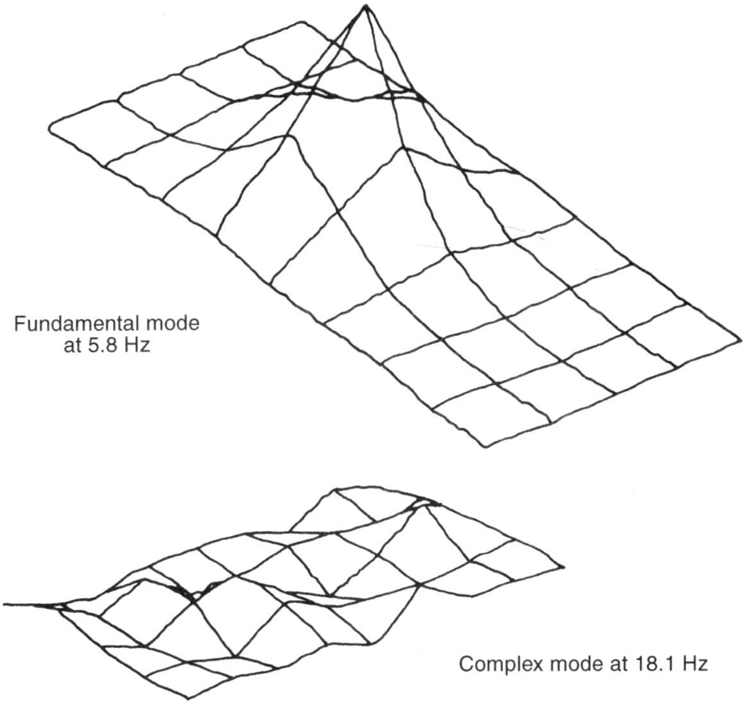

Fig. 6.70 Measured mode shape at Middlesbrough floor, with an added mass.

conducted on the building, and at that time the author conducted induced vibration tests on the building in question.

It was clear to see from those tests that the joints in the building were not performing satisfactorily in some cases. Figure 6.72 shows the results of measurements made whilst shaking the structure in a torsional mode. The sharp differences in the response of some adjacent elements is an indication that the joints between those elements were not performing satisfactorily. Indeed, other tests on the building suggested that it behaved rather like a giant tuning fork, with little connection between the two halves of the building above the eleventh floor.

The building was one of a series constructed to the Larsen Neilsen design. The design had been produced in Denmark, for buildings of not more than five storeys in height. The system had been prohibited completely in the USA. Nevertheless, in the UK, construction of blocks of up to 22 storeys in height had occurred on a large scale.

The main problem was one of 'buildability'. As construction had progressed, the joints between adjacent large panels had accumulated rubbish, which had not been cleared when grouting. The joints had therefore been suspect from the very beginning.

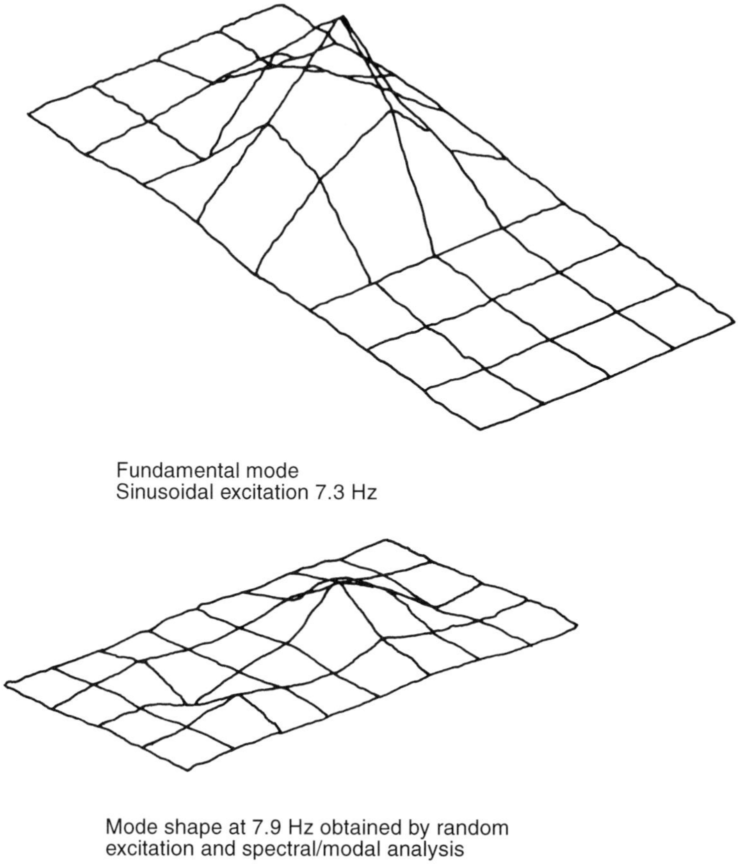

Fig. 6.71 Measured mode shape at Middlesbrough floor, with no added mass.

As a result of tests carried out on the building, and a great deal of adverse publicity, the building was demolished.

An experiment on a one-quarter-scale model of a Larsen Neilsen type of design was conducted (Fig. 6.74), and dynamic and static load tests were conducted on the model, which was designed to a new improved code format. Unfortunately, the model also demonstrated a progressive collapse under a simulated gas explosion, notwithstanding the increased tolerance built in to the new design. The conclusion of some studies of the explosive gas dynamics was that a resonance condition inside a room of the building may have caused an additional dynamic magnification factor that was not previously considered. The progressive collapse of the model is shown in Fig. 6.75.

Interestingly, the attempts to clear away a very badly damaged building from a laboratory demonstrated just how much redundancy there is in the

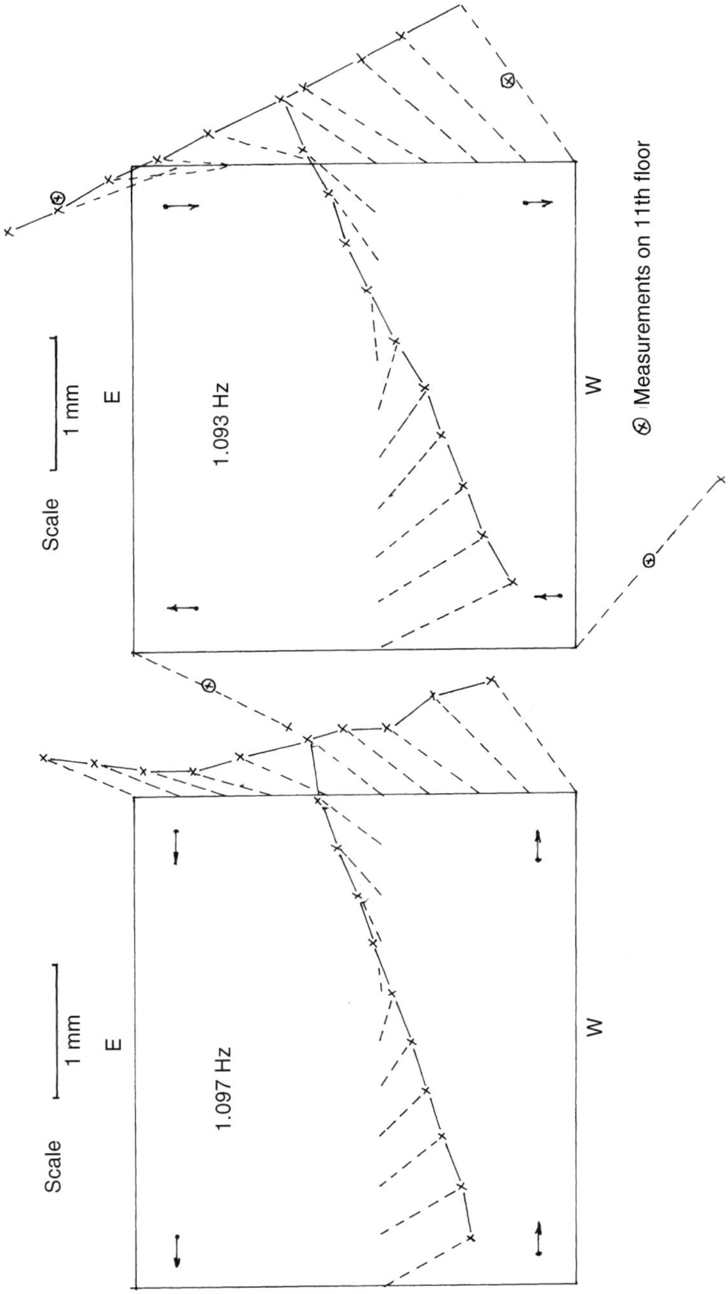

Fig. 6.72 First torsional mode – plan displacements, Ronan Point.

Fig. 6.73 Ronan Point, collapse after gas explosion.

Fig. 6.74 One-quarter-scale model of a Larsen Neilsen type of building.

Fig. 6.75 Progressive collapse of model of Ronan Point.

strength of a tall building. Indeed, the fact that the original Ronan Point building stood for so long, despite the fact that it was in a poor state of health, is testimony to the reserves of strength that often appear in structures.

CHAPTER 7 *Damping*

It has been emphasized throughout the preceding chapters how the damping value of a mode of vibration governs the overall response in that mode of vibration. In the preceding chapters, it has been tacitly assumed that the damping value is constant and that it is based on a mechanism that is known not to be present in structures. In this chapter, the real situation, in which damping changes with amplitude, is explored. The non-linearities involved, whilst not large in absolute terms, make a large difference to the response of a structure. It is quite normal for damping to change by 100% of its value over the amplitude range involved with the response to wind or wave, whilst for earthquakes the value may change by up to 500%. This factor is perhaps rather significant in the conventional differences between such fields of study, and at the same time it emphasizes why they are merely sub-branches of the same systematic approach. Since the response of a structure is inversely proportional to the value of damping, such large differences are the dominant influence in the calculation of the response of a structure.

As has previously been observed, the damping value of a mode of vibration is a quantification of the amount of energy loss when the structure vibrates in that mode. There are several different ways of describing this energy loss, although all of the commonly used expressions are equivalent. Throughout this book, damping is expressed as a percentage of the critical value. The equations developed in Chapter 3 give the rationale for this. It is as well, then, to start this section with a number of definitions that will clarify the use of the damping parameter.

- Critical damping is that quantity of energy loss in a mode of vibration that causes a system just to fail to oscillate. This condition is known in some fields as a 'dead-beat' response.
- Zeta (ζ), the damping ratio, is the actual quantity of damping in a mode of vibration divided by the critical value for that mode.
- Delta (δ), the logarithmic decrement, is the natural logarithm of the decrement per complete cycle of oscillation when a system is vibrating in a mode of vibration; δ can be related to ζ by the following expression:

$$\delta = \frac{2\pi\zeta}{\sqrt{(1-\zeta^2)}} \tag{7.1}$$

All of these definitions refer to damping that is assumed to have a mechanism of viscous damping. In the development of the equations of motion in Chapter 3, it was assumed that the damping force is proportional to velocity.

This refers to the loss of energy in fluids that occurs when their viscosity causes energy to be lost. To get an intuitive impression of this mechanism, the reader is invited to imagine trying to walk through treacle. A commonly used device that actually uses this mechanism is a damper on an automobile. The dashpot device is set up to cause a piston to try to displace a fluid. In a structure, such a physical mechanism does not normally exist.

Figure 7.1 shows a decay of oscillation constructed assuming that the viscous mechanism exists. As can be seen, the envelope of the decay is a perfect exponential decay.

In practice, when structures are shaken and then allowed to oscillate freely, they do not show such a characteristic. Figure 7.2 shows the envelope of the decay function of a 20-storey tall building. It can be seen that the

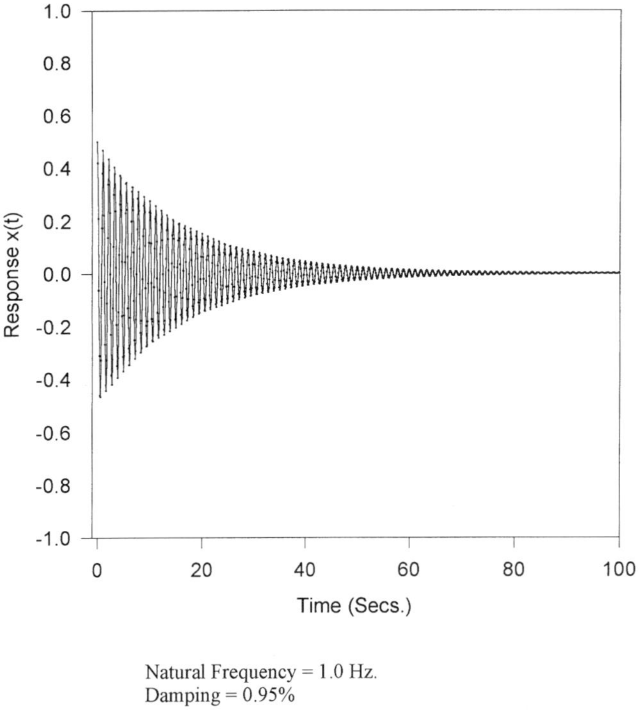

Fig. 7.1 Decay of oscillation: natural frequency = 1.0 Hz; damping = 0.95%.

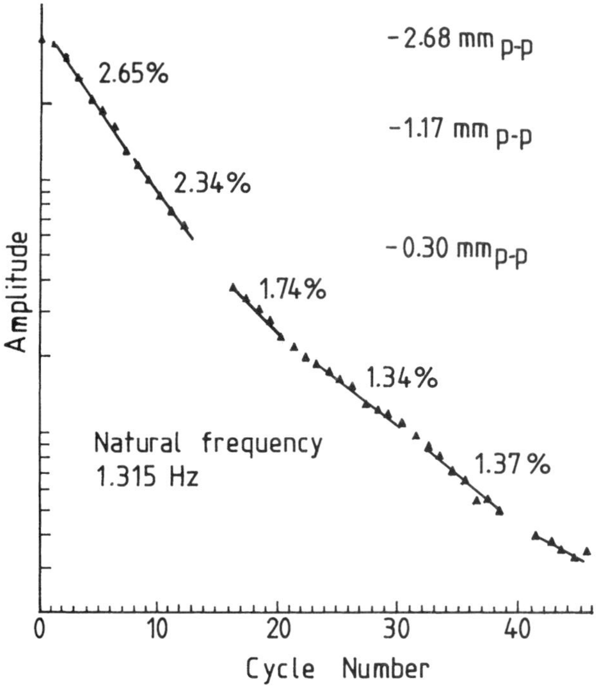

Fig. 7.2 Extended decay of oscillation: natural frequency = 1.315 Hz.

value of damping decreases from 2.65% to about 1.35%. When such decays of oscillation first appeared, the suspicion arose that there was a 'low-amplitude plateau' region of damping.

Early in the studies of the damping behaviour of structures, the concept of an 'equivalent viscous damping value' was introduced. This allowed the use of convenient mathematical equations, which assume that such a mechanism exists despite the fact that the mechanism was clearly not operating. Whilst this simplified many aspects of the calculation of dynamic behaviour, it also left a legacy of misunderstanding, and in particular it was an invitation to forget that damping is non-linear in the vast majority of cases. Luckily, some recent work by Wyatt (1977) has shown that a multitude of friction mechanisms, which are clearly present in structures, when summed over a range of amplitudes, actually produce a result that looks like a viscous-type response. Truly, we now have the complete physical picture for the implementation of the equivalent viscous damping approach, which also happens to explain why non-linearities should be expected.

With this understanding, there is a full justification for using the equivalent viscous damping approach, so long as the approach is not extended out of a suitably small range of amplitudes in which the damping value can be considered to be constant. If calculating the response of a structure at a single amplitude (say the design response), then this condition is clearly met.

If extension outside a small range is required, there is also an approach that will describe the way in which the equivalent viscous damping value changes with amplitude.

Wyatt's work with footbridges led to his production of a model of damping behaviour that included the concept of stuck elements. In such cases, a certain level of force is necessary to overcome the attraction between the two elements. At such a stage, the two elements will begin to move relative to each other, and will dissipate energy through the rubbing surfaces. The greater the 'stuckness', the larger the initial force necessary to produce relative motion between the elements.

Herein lies the solution to the problem of a description of damping behaviour, because the amount of 'stuckness' is well defined in the field of fracture mechanics. The longer the boundary between two elements, the lower the force necessary to produce relative motion. With two small elements requiring a large force to separate them, we are in the ambit of intermolecular forces. Whilst at first sight this may not seem important to structural engineers, in practice the opposite is the case. Cracking in the material of the structure starts at a molecular level, and the initiation of cracking is now seen to be the key to the start of relative motion between elements in a structure. The forces that initiate cracking in a structure, or that cause relative motion between large elements (such as beams and columns), are shear forces themselves caused by the dynamic motion of the structure. This is another unifying theme, because the same mechanism that leads to a degradation of stiffness of a structure also causes there to be a greater amount of damping.

It is also worth noting that micro-surfaces are a function of the structure itself rather than any individual mode of vibration; yet, despite this, there are different amounts of damping in different modes of vibration for the same structure. This situation is caused by the different distortions in the fabric of the structure in different modes. Reference to the typical mode shapes shown in Chapter 4 will give an intuitive understanding of this comment.

7.1 Fracture mechanics and energy dissipation in structures

The early work by Griffith (1920) that started the entire study of fracture mechanics was an attempt to explain the difference between the theoretical maximum strength of materials and their significantly reduced practical levels. Griffiths' explanation relied on a description of the conditions necessary for a crack to elongate under a remote stress field. His proof that there is effectively a greatly increased stress intensity in the region of a crack tip led to a better understanding of the in-service performance of materials.

More recently, the Griffiths approach has been extended somewhat, although the essential ideas remain intact. Considerations now involve three types of mechanism for crack elongation, depending on whether shear, tension or a combination is acting on an imperfection in the material; but, more significantly, a fracture process zone ahead of a crack is also considered.

There are essentially three mechanisms to consider for energy dissipation in the presence of crack (or micro-imperfection) extension:

- Surface energy dissipation caused by extension of existing surfaces.
- Energy dissipation through the formation of micro-cracks in the fracture process zone ahead of a crack or micro-imperfection.
- Energy dissipation in the wake of a crack.

This third type of energy dissipation appears to be significant for concrete, and has been termed 'aggregate interlock'. Such energy dissipation can be calculated from cohesive crack models in the case of small-scale mechanisms.

There are two basic types of mechanism for crack extension mechanisms. These are termed mode I (which is an opening type) and mode II (which is a sliding type). A further type, mode III, which is termed a mixed mode, is also considered. The critical concept for these models is the stress intensity factor. For these three modes, these are respectively K_{I}, K_{II}, and K_{c}.

It is necessary now to direct the discussion towards the effect of such considerations in conjunction with the observed behaviour of entire structures.

As has been observed by authors investigating large-scale cracking in concrete, there are three main regimes for fracture in concrete:

- Under small initial loads, the material behaves essentially elastically.
- Under increasing load, a zone of micro-cracking in the fracture process zone rapidly occurs; however, main crack growth in the form of elongation of cracks or imperfections does not occur.
- Main crack growth occurs.

These three mechanisms at the micro-scale integrate to a holistic effect that is observable at the macro-scale:

1 At very low amplitude, the structure dissipates very little energy and moves quickly to an amplitude at which an energy balance is struck. This occurs when large structural elements move relative to each other and occurs at small forces. This area has previously been termed the 'zero-amplitude damping'. Figure 7.3 shows the type of elements that are involved in this type of energy dissipation. The example is taken from the Petronas Towers whilst it was under construction.

Fig. 7.3 Petronas Towers during construction, showing interaction of large structural elements.

2 At increasing amplitude, energy is dissipated in the material of the structure, through the mechanism of micro-cracking. This can be self-repairing. This area involves a non-linear damping characteristic. Figures 7.4–7.6 show areas that contribute to the non-linear damping zone.

Fig. 7.4 Joints of steel and concrete frames.

Fig. 7.5 Joints in a small structure with a wooden roof.

Fig. 7.6 Attachment of cladding to a floor.

Figure 7.4 shows areas of interaction between intermediate-sized elements, whilst a smaller connection between a column and a roof also contributes to this damping zone (Fig. 7.5). Figure 7.6 shows the attachment details of cladding panels. Movement of such joints also contributes to the non-linear energy dissipation area. At a higher amplitude, imperfections in the material of the structure are made to work. Figure 7.7 shows such imperfections in a concrete sample.

Fig. 7.7 Concrete wall with visible imperfection.

3 At amplitudes large enough to mobilize all mechanisms, and to allow main crack growth, a new energy balance is struck. This is 'high-amplitude plateau damping', and is typically observed under earthquake loading.

In Fig. 7.8 the generalized damping characteristic is depicted.

Finally, it should be observed that there is an additional mechanism that occurs at very high loading, in which damage occurs in major cracks, friction drops and accordingly the high-amplitude plateau falls. This mechanism is seen only when large-amplitude motion continues for some time, as has occurred for instance in one large-amplitude induced vibration test by Galambos and Mayes (1979). For most circumstances, this is not a practical consideration, since we are dealing with force levels that would be associated with major earthquakes, and these rarely continue for long enough to activate such a type of behaviour. In practice, then, there are two regimes that are important to the study of damping in structures. The first involves the relative motion of large elements in the structure, and the second involves the initiation of cracking in the material of the structure (and of 'non-structural' elements).

At extremely small amplitudes of motion, structural elements dominate the damping characteristic. In fact, at amplitudes that are less than that necessary to bring such elements into relative motion, there will be little damping available. With no energy dissipation mechanism available, a minute force causes the structural motion to rise quickly to the point at which an energy balance is struck. The amount of damping at this point is

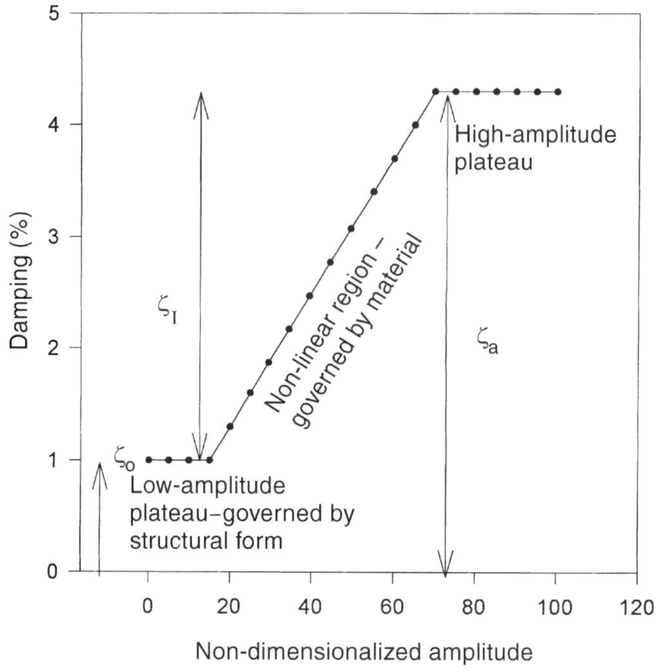

Fig. 7.8 Generalized damping characteristic for large structures.

termed the 'zero-amplitude damping'. Quite clearly the semantics are difficult here, since there can be no damping at zero amplitude, and more correctly the term should be 'extremely-low-amplitude damping'. However, the term is useful, and measurement of damping behaviour at those extremely low amplitudes has so far proved to be an intractable problem.

There then follows an amplitude regime in which there is an absence of elements of the threshold length in the structure. Measurements made so far on full-scale structures strongly suggest that this region extends from about 0.5 m down to about 20 mm. This region is termed the 'elemental gap' and is associated with a constant damping characteristic.

The largest cracks that participate in energy dissipation in a structure seem to be in the region of 20 mm long. As shear forces increase, more and more cracks are induced to increase their length as they are torn. This situation is described in detail in the study of fracture mechanics. As the amplitude of motion increases, smaller and smaller cracks in the material of the structure are mobilized until an effective limit is reached at approximately 100 μm. Since there are large numbers of potential cracks, and elongating cracks within the material of a completed structure, then there is great potential for energy dissipation. This region of motion is the one in which damping increases with increasing amplitude of motion of the structure.

Effectively, then, this amplitude region, in which damping increases, occurs when cracks have initial lengths in the range 10^{-3} to 10^{-8} m. This implies, in turn, that damping will increase only over a region of approximately three orders of magnitude of the amplitude of the response of the structure.

Finally, there occurs a region in which the amplitude of motion is so large that all possible cracks are mobilized, and the damping value remains constant. This coincides with the experience of earthquake engineers. When a structure responds to some large force in this manner, and when this motion occurs over an extended period, then the friction force (which causes the energy dissipation) between elements will tend to decrease with the passing of time, since large-scale rough surfaces will tend to become smooth, and the potential for energy dissipation will reduce. In the only test to have been conducted at high amplitude over a long period, such a characteristic was observed. This work is reported by Galambos and Mayes (1979).

The suspicion that a low-amplitude plateau existed was investigated by observing many values of damping that could be guaranteed to occur at very low amplitude. The values obtained suggested that the low-amplitude value of damping (ζ_0) is correlated with the numerical value of the fundamental natural frequency.

Similar relationships appear to hold for all structures, and as a rough approximation the low-amplitude plateau value of damping can be taken as numerically equal to the fundamental natural frequency. That is, if the natural frequency is 0.8 Hz, then the best-guess value for ζ_0 is 0.8%.

Data for the rate of increase in damping were also assembled. When these were plotted, Fig. 7.9 was the result. The implication is that the rate of change of damping is proportional to a base dimension of a structure. Considering the mechanism discussed above, this is a logical conclusion. The more material available, then the greater the rate of increase in damping.

Very recently, techniques have appeared that allow the measurement of damping from the random excitation of structures at different amplitudes. Figure 7.10 is a diagram of such a damping measurement taken from a very tall building. The three elements of the generalized curve are clearly visible, and are a vindication of the theory outlined above. Of particular interest is the appearance of the low-amplitude plateau, which has proved extremely difficult to obtain but appears here for the first time.

7.2 A mathematical approach that is consistent with codification

The estimation of damping values for a structure in the absence of measurement is still somewhat problematical. However, based on the limited measurements that have been made to date under controlled conditions, the

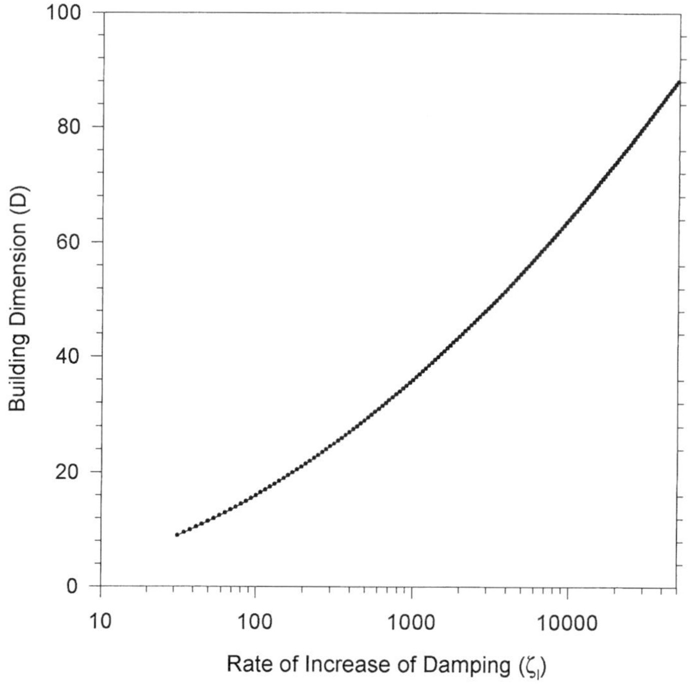

Fig. 7.9 Damping increase versus building dimension characteristic.

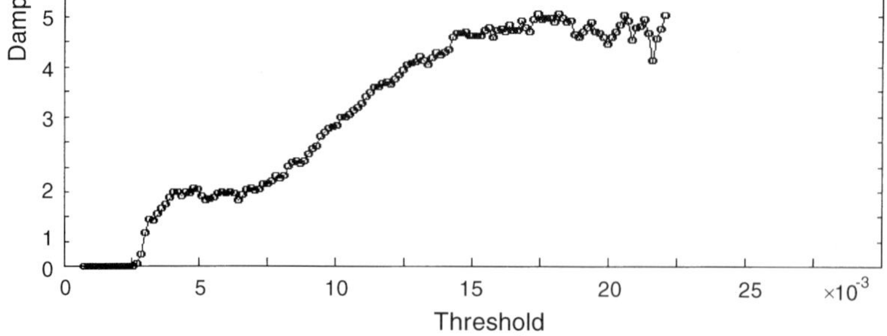

Fig. 7.10 Typical damping curve from a building.

parameter can be estimated for a wide range of structures in the first two sections of the damping regime. This includes the 'zero-amplitude' value and the non-linear zone. There is additionally an indication of the effective limit to the non-linear zone. Luckily, this covers all of the situations likely to occur in the study of dynamics of structures at present.

The expression used to describe the equivalent viscous damping value for a single mode of vibration at different amplitudes of response is

$$\zeta_a = \zeta_0 + \zeta_{\mathrm{I}} \tag{7.2}$$

where ζ_0 is the 'zero-amplitude damping value' and ζ_{I} is the 'damping increment'.

In turn, ζ_0 and ζ_{I} can be calculated from algebraic-type equations for different types of structure. The quantity and quality of data available for the reliable estimation of these parameters is variable. However, certain basic principles seem to apply in several different fields. Regression studies of very-low-amplitude data suggest a series of different formulae as being possible, but because there are so few data on which to base the estimate it is sufficient to simplify to the following:

$$\zeta_0 = 0.15 + J_n(f_n) \tag{7.3}$$

where f_0 is the natural frequency of the mode of vibration under consideration and J is a multiplying factor, which currently is estimated to be 0.76 for $n = 1$.

The value of ζ_I has been estimated previously for the fundamental mode of vibration for tall structures, and the following is a simple method for calculation:

$$\log_{10} \zeta_{\mathrm{I}} = \sqrt{(D/2)} \tag{7.4}$$

where D is the dimension of the structure in the direction of motion of the mode of vibration.

The entire formula for the calculation of a damping value at amplitude x for a tall structure then becomes:

$$\zeta_x = 0.15 + J_n(f_n) + \frac{x\zeta_{\mathrm{I}}}{10H} \tag{7.5}$$

where x is the peak displacement (in mm), H is the height of the building (in metres) and ζ_{I} is the rate of increase of damping taken from Fig. 7.9. Thus for a 100 m tall building with base dimension 20 m, ζ_{I} takes a value of 172. If the design amplitude is 10 mm peak deflection, then ζ_x can be estimated as follows:

$$f_n = 46/100 = 0.46\,\mathrm{Hz}$$
$$J_n = 1 \quad \text{(for the fundamental mode)}$$

and thus

$$\zeta_x = 0.15 + (1 \times 0.46) + \frac{10 \times 172}{10 \times 100} = 2.33\% \tag{7.5A}$$

For much taller buildings, there are indications that the upper plateau is reached quickly and that the equation above may need to be modified. This is because the range of amplitudes over which the non-linear zone operates is approximately 3.5 decades. If the break point from the low-amplitude plateau is established, then this amplitude range can be assessed. This is discussed further below.

It is worth observing here that, if D is small, then the damping value will be essentially constant across a large range of amplitudes. This explains why, for instance, tall chimneys show little increase in the damping value over large ranges of amplitude.

Information gathered from several different types of structure tends to suggest that the relationship between zero-amplitude damping and fundamental natural frequency holds good for all structures. Whether or not this continues to be true as more data are added, the relationship is an extremely useful starting point for the estimation of likely damping values.

It is possible to estimate values of ζ for other types of structure and for other, more distorted, modes of vibration. Table 7.1 towards the end of this chapter gives details.

Figure 7.11 shows a distillation of information coming from concrete-frame and steel-frame structures. The suggestion is that the low-amplitude

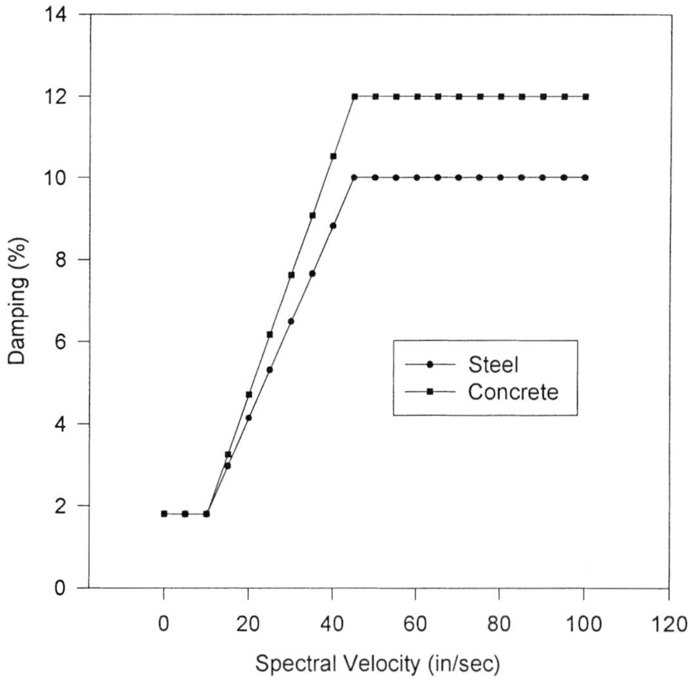

Fig. 7.11 Damping versus magnitude of spectral velocity.

plateau, being caused by a mechanism that is not material-dependent, starts at a similar value for similar types of structural form. The rate of increase is slightly greater if concrete is used, and results in a high-amplitude plateau that is approximately 20% greater.

7.2.1 Detailed consideration of the non-linear zone

The surface energy in a crack can be represented as

$$\gamma = \frac{K_{\mathrm{Ic}}^2}{2E(1-\mu)^2} \tag{7.6}$$

Also the length of an imperfection that is mobilized by a stress (σ) is given by

$$l = \frac{2\gamma E}{\pi\sigma_1^2} \tag{7.7}$$

This implies that the relationship between the stress appearing in the imperfection and the length of crack just starting to extend is

$$\sqrt{\frac{l_{\mathrm{h}}}{l_{\mathrm{l}}}} = \frac{\sigma_{\mathrm{l}}}{\sigma_{\mathrm{h}}} \tag{7.8}$$

Measurements on both steel and concrete tend to suggest that, within the material, there are likely to exist cracks of the order of millimetres in length, whilst studies of concrete under a scanning electron microscope suggest that the lower limit is 100 nm (Fig. 7.12). If we assume the largest material imperfection to be 20 mm, then this implies in turn that the ratio of the force just causing the largest material imperfections to elongate to the force that just makes the smallest imperfection elongate is 1414.

The ratio of the in-crack stress to the far-field stress varies with the magnitude of the far stress. This is because the ability of the force to appear on the surface of the imperfection is more efficient when the crack length is greater (at smaller forces).

Work by Dharan and Recker (1991) has demonstrated a factor of 1.6 for crack lengths of 9 mm down to 0.09 mm. However, the ratio was constant from 9 mm down to 1.8 mm. This merely shows that the factor is non-linear, and in any case it has not been observed for cracks in the nanometre range. There is also evidence that the critical stress intensity factor (K_{Ic}) changes with amplitude. Again, measurements down to the smallest crack lengths do not exist.

Clearly, then, there is a ratio between the critical far-field force that just initiates the start of the non-linear region, and the force that takes the damping characteristic onto the high-amplitude plateau, which can be expressed as

$$F_{\mathrm{h}} = F_{\mathrm{l}} \times 1414 \times K \tag{7.9}$$

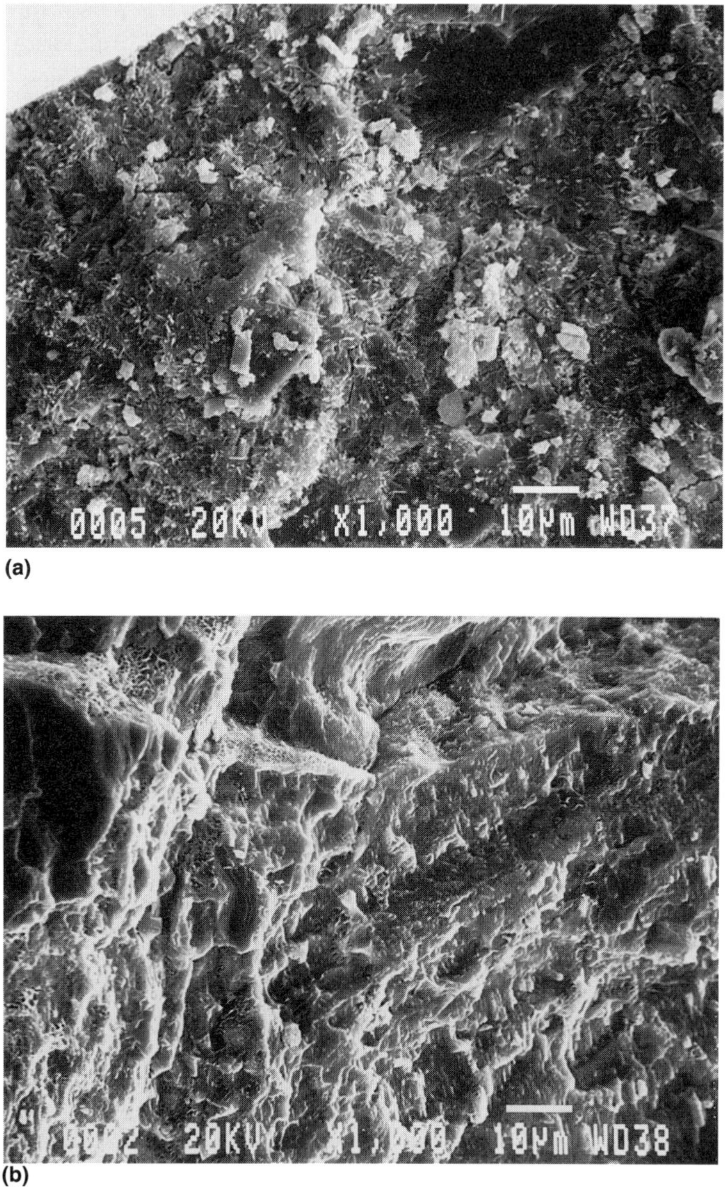

Fig. 7.12 Scanning electron microscope images of (a) OPC concrete and (b) PFA concrete.

where *K* represents the change in the ratios of the far-field force to the in-crack stress as modified by the change in critical stress intensity factor over the amplitude range 2 mm to 100 nm.

Arbitrarily, the value of K is assigned a value of 7 in the following section. It will be established by experiment in any forced vibration test that takes a structure to the high-amplitude plateau.

7.3 *Example from John Russell Court*

Figure 7.13 is taken from the testing of the John Russell Court building, Edinburgh. This diagram is the result of inducing vibration in the structure with a vibrator system and then switching off the excitation. The resulting decay of oscillation starts from a well defined force and amplitude, and allows the identification of the low-amplitude break point at the initiation of the low-amplitude plateau. As can be seen from the figure, this occurs at a force of approximately 700 N, and a corresponding amplitude of 0.15 mm. The low-amplitude plateau region can be clearly observed after this break point. The implication of this is that

$$F_l = 700\,\mathrm{N} \tag{7.10}$$

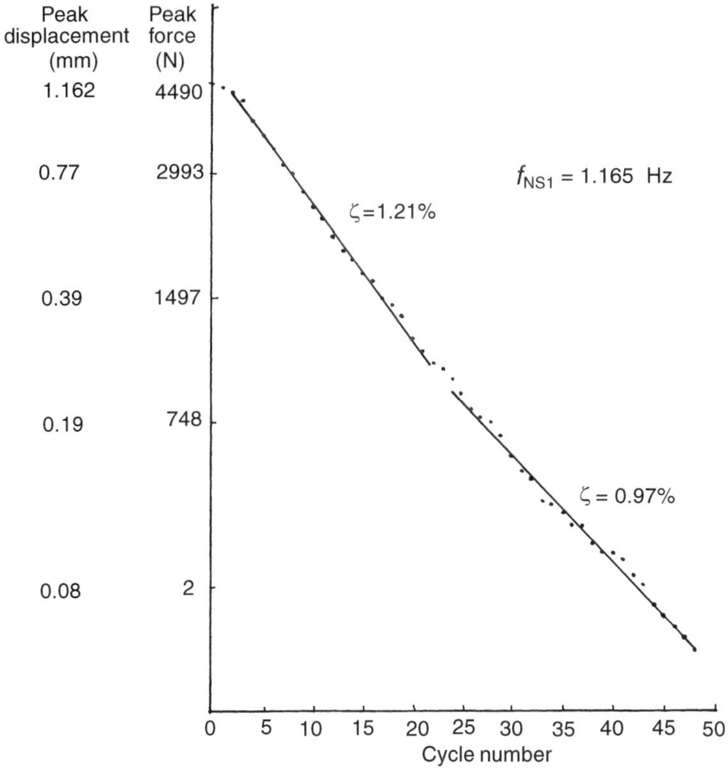

Fig. 7.13 John Russell Court, decay of oscillation.

In turn this implies that

$$F_\mathrm{h} = 700 \times 1414 \times 7 \tag{7.11}$$

It is then possible to take the equation of motion for the resonance condition of a single mode of vibration:

$$A = \frac{F_r}{M_r 8 \zeta_A \pi^2 f_r^2} \tag{7.12}$$

and substitute the measured damping non-linearity.

The damping characteristic can be expressed as

$$\zeta_A = \zeta_0 + \zeta_\mathrm{I} \tag{7.13}$$

where A is the displacement at the top of the building in metres, and damping is expressed as a percentage of critical damping at amplitude A. For the case of the John Russell Court building, the measured damping characteristic was

$$\zeta_A = 1.08 + 290A \tag{7.14}$$

If this expression is substituted into equation (7.3), then a relationship between the non-linear damping value and the amplitude of motion at the top of the building is obtained as follows:

$$A = \frac{-\zeta_0 + \sqrt{\zeta_0^2 + (4\zeta_I F_r / M_r 8\pi^2 f_r^2)}}{2\zeta_I} \tag{7.15}$$

Substituting the expression for the non-linear damping characteristic in this equation, then it is found that the high-amplitude plateau would be reached at a peak amplitude of 7.1 mm, at which the damping ratio would be 3.13%.

It is interesting to note that the modal force in this case is 7.08 MN. This compares with a value for 10 m 'fifty-year return period' winds which are often in the region of 1 MN.

7.4 Other energy-dissipation mechanisms

Friction within the fabric of a structure is normally the most important mechanism for energy dissipation. However, there are important exceptions. Principal amongst these are the mechanisms of soil-structure interaction and of artificially added damping. These are considered in turn.

7.4.1 Soil-structure interaction

Large structures can, and sometimes do, cause the surrounding terrain to move. Waves moving away from structures that have been shaken in controlled testing have been measured and quantified. This clearly represents a major source of energy dissipation, and when present can cause large differences to the amount of damping present. Once again, however, this effect is not of great significance for the majority of large structures in their fundamental modes of vibration. However, even for very large structures, the effect can be large for more complex modes of vibration.

7.4.2 Artificial damping mechanisms

Perhaps the best known of the damping systems installed in large structures is the damping system in the World Trade Center in New York City. In this system, a distortion induced in a viscoelastic material appears to dissipate a significant amount of energy. The activation of the distortion of the devices is caused by a small distortion that occurs between the wall and the floor at each floor level. Inspection of the devices after more than a decade of continuous use has shown there to be insignificant degradation of the material.

The World Trade Center was one of the early attempts in a commercially operating building to reduce the cost of materials whilst at the same time providing security from the inconvenient motion produced even in the presence of quite small forces. The economic benefits of such an approach are such that many others have attempted to introduce systems to prevent motion that will be perceptible in inhabited structures.

Passive systems, such as that in the World Trade Center, merely add to the overall energy dissipation. The prize of a system that would actively reduce motion has also been pursued. A major requirement for such systems would be to reduce motion at times of severe wind storms or during earthquakes. Unfortunately, these times also happen to carry a high risk of power failure. In such a situation, a non-operational active damping system would actually act as a prejudice to the structure.

There have been two approaches to this problem to date. One is to implement a system that is self-powered or stores its own power, and the other is to supply a local source of power that can be made to have a very short start-up time.

For non-extreme situations, it appears that the technology to reduce motion to very small quantities already exists. To date, there is no evidence as to how such systems behave in extreme situations.

The value of damping to be used for design purposes is derived from the formula:

$$\zeta = \zeta_0 + \zeta_{\mathrm{I}} + \zeta_z \tag{7.16}$$

The values for ζ_0 and ζ_{I} are taken from Table 7.1 unless more direct evidence for the value of ζ exists. The value of ζ_z should be taken as zero unless specific information is available to the particular design case. The value of ζ_z derives from a consideration of one of the following forms of damping:

- Aerodynamic damping (A)
- Radiation damping (R)
- Artificially induced damping (I)

Aerodynamic damping will exist where there is a coupling between the motion of the structure and the flow of the wind. Radiation damping exists when there is significant soil-structure interaction. Artificially induced damping may be as a result of the use of passive damping systems, such as viscoelastic rubber dampers, or by the use of active damping systems. Such systems may generate antiphase motion through reaction or may dissipate energy directly into liquids or other energy sinks. For account to be taken of active damping systems, the system must operate even in the event of power failure.

It should be noted that the values for damping quoted in Table 7.1 are the author's recommendations based on measurements, and extrapolations from those measurements. Nevertheless, it must be pointed out that there is a small database available to support the recommendations of Table 7.1.

Table 7.1 Values for ζ_0, ζ_I and ζ_z

Structure	ζ_0 *Any material*	ζ_I *Concrete*	ζ_I *Steel*	ζ_z
Tall building	$46/H$	ζ_0	$0.8\ \zeta_0$	A + R + I
Chimney	$46/H$	$\zeta_0/10$	$\zeta_0/20$	A + I
Bridge	$800L_s^{-0.9}$	ζ_0	$0.8\ \zeta_0$	A + I

Key:
A Aerodynamic damping
R Radiation damping
I Artificially induced damping

CHAPTER 8 *Synthesis*

In this chapter, examples are given of the approach to the estimation of the response of a structure when an extreme event occurs. This calculation can be performed at the time of design of the structure, since all parameters can be estimated without any measurement of the actual behaviour of the structure. It should be remembered, however, that in the absence of measurements from the structure there is a confidence interval associated with the estimate of the response. The size of this interval will be dependent on the quality of the information leading to the estimate.

The first stage in design is to decide what criteria will apply. The norms are as follows:

- Tall buildings – comfort
- Bridges – serviceability
- Chimneys – serviceability
- Dams – survivability
- Floors – serviceability
- Offshore structures – serviceability

There are many codes of practice that specify acceptable vibration limits. In general, these are based on early work in Germany that led to the initiation of the DIN 4150 standard. Essentially, this sets limits to vibration input at 15 mm/s as the acceptable criterion. BS 5228 (Part 4) also assumes a peak particle velocity (ppv) of 15 mm/s as the maximum acceptable.

More recent information suggests that the damage initiation curve has a break point at 5 Hz. CEB Task Group V4 reviewed all sources in 1988, and suggested that below 5 Hz the damage initiation characteristic was proportional to acceleration rather than to velocity. Table 5.1 of that document quotes a value of ppv of 12 mm/s for industrial buildings of reinforced concrete, but notes that this applies to the range 10–30 Hz.

The US Bureau of Mines reported in 1980 (RI 8507, US Department of the Interior) that the DIN 4150 standard, which had been used as a reference since its inception in 1939, was unworkable. At the same time it reported the work of Esteves (1978) and his recommendation that for concrete structures a value of 15 mm/s should be used.

The International Standards Organization Technical Committee 98 reviewed the subject, and in 1988 produced a draft international standard. This later became ISO 2631. Once again, however, the data related only to structures with natural frequencies under 10 Hz. The acceptable levels quoted were of 5–10 mm/s for residential structures and 20 mm/s for industrial structures.

All of these criteria represent the *input* to the structure, and as such they can be only loosely correlated with damage criteria. The actual damage will be a function of the structural form, the influence function for soil–structure interaction, and the stress concentrations within the structure.

The DIN 4150 criteria were closely related to the kinetic energy in the vibration, whilst the US Bureau of Mines data mainly came from measurements on American domestic housing close to a surface mining operation. Figure 8.1 is redrawn from the US Bureau of Mines data, and shows a summary of all tests reported in 1980 after a major study. There are two factors of particular importance here. Firstly, the minor damage line corresponds with the recommendations of most codes of practice. Secondly, the few data below 3 Hz, and the increased variance, emphasize this author's cautionary remarks about extrapolating to very low frequencies.

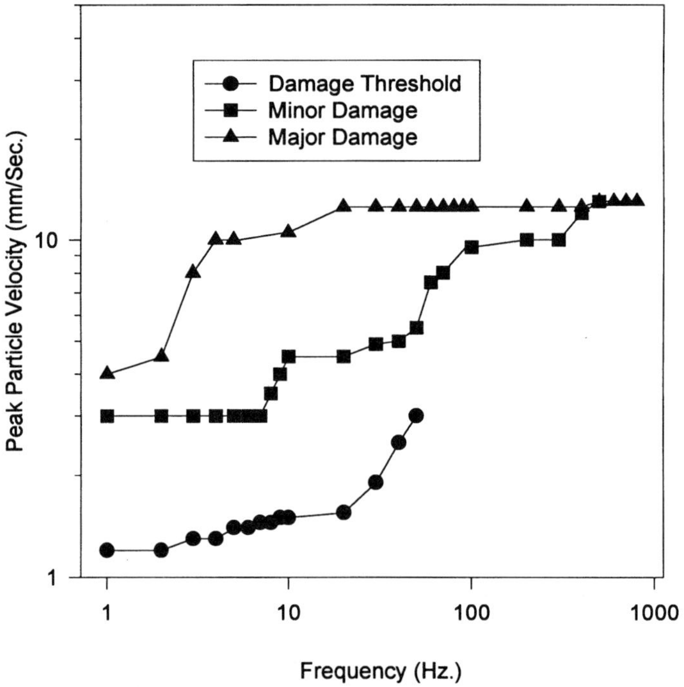

Fig. 8.1 Damage as a function of peak particle velocity.

Having ascertained the performance required from Fig. 8.1, it is necessary to decide what is likely to be the action that will produce the design loading. The approach to design is then dictated. The main sources are considered below.

8.1 Wind excitation

Because the wind represents a random source, then it is normal to use a spectrally based approach to design. However, the interaction of the wind with the structure can be quite complex, and modifications to the forcing function spectrum are introduced, through such functions as the aerodynamic admittance and the background spectrum. A major disadvantage of this approach is that phase information is disregarded. However, at the time of writing, it represents the current state-of-the-art approach.

8.1.1 Design strategy for wind engineering

The response of a building or other structure in a strong wind comprises two parts:

- Quasi-static response
- Dynamic response

The quasi-static response is the averaged mean 'lean' of a building (or other structure) in response to the force of the wind moving it in the direction of the force.

The dynamic response of a building or other structure occurs at preferred or natural frequencies of the entire structure. These natural frequencies represent the inverse of the time taken to complete a cycle of oscillation in one of the natural modes of vibration of the structure. Each mode of vibration (or deflected form) has an associated natural frequency. When a building or other structure is induced to respond at these natural frequencies, then it is said to be in a state of resonance. Resonant response accounts for a large part of the overall response of a structure when it is subjected to large dynamic forces such as those which occur in typhoons or earthquakes. The dominant resonance of tall buildings takes a deflected mode shape, which is similar to that of a ruler that has its end 'twanged'.

In order to calculate the probable response of a structure, the wind or earthquake forces have to be described in statistical terms, since they are by definition unsteady forces. To effect this, we normally deal with forces that occur on average only once during a time period. The normal length of time considered for design against typhoons (and indeed the value that is used in the Hong Kong code of practice on wind effects on buildings) is a 'fifty-year wind'. There is considerable misunderstanding of this term. The actual definition is as follows:

Fifty-year wind: that wind which occurs, on average, just once every 50 years.

In practice, it would be better to consider the inverse relationship:

Fifty-year wind: that wind which has a yearly probability of occurrence of 0.02.

This would overcome the problem of giving the impression that the 'fifty-year wind' will occur only after 50 years. It could occur today.

8.2 Earthquake excitation

For the case of earthquake loading, we again face a random function for the source, but in this case the duration is so short that true spectral analysis cannot be performed because spectra do not become stable in the times available. The approach, then, is to adopt the spectral approach to the earthquake situation, or to use a time stepping function to calculate real-time response.

In the case that a pseudo-spectral approach is used, then design spectra for a site can be assembled and averaged, or 'typical' design spectra can be used. Simply because such events have not been observed often enough to present a stable statistical definition of the excitation, the approach is subject to a wide confidence interval. However, this approach is often used in codification.

If a time stepping calculation is used, it is still necessary to assemble a 'typical' acceleration history for an earthquake in the region being considered. Standard earthquake time histories can be used, and it is even possible to try several different standard time histories as input for a calculation of the time history of a particular structure. The worst case is then taken for design purposes.

8.3 Traffic excitation

Whilst the input from traffic is a random process, the input from individual vehicles can have distinctive spectral characteristics. Information from Swiss tests has shown the characteristics to take the form of Fig. 8.2 for lorries.

8.4 Human excitation

As has been observed previously, the human body has a tolerance to the speed of walking, and as such jumping occurs mainly at a frequency centred on 2 Hz. However, it is possible for people to synchronize to frequencies as high as 5.5 Hz. In the case of jump dancing (pogo dancing) being the excitation source for a floor, a special approach that is deterministic is possible.

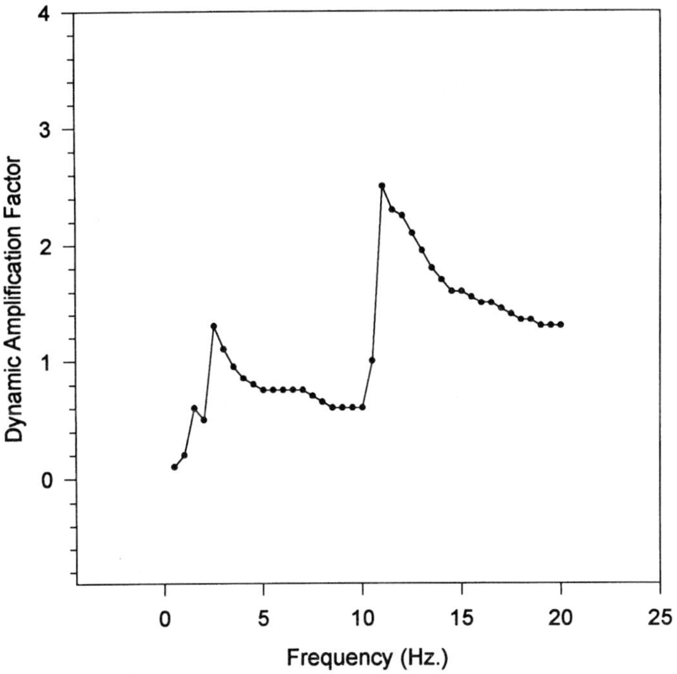

Fig. 8.2 Dynamic amplification resulting from a lorry driving over a 4.5 mm discontinuity.

8.5 Machine excitation

In this case, the excitation source is basically sinusoidal and deterministic, and so a straightforward algebraic calculation method is recommended.

In the case of each type of action on a structure, the steps necessary to calculate the response are similar if the systems approach is adopted. However, the details vary quite considerably, depending on the type of action that occurs.

8.6 Generalized approach to design

The approach in each case is as follows:

1. Assemble data about the occurrence of the action.
2. Decide on the criterion for design (e.g. comfort or survivability).
3. Make an assessment of the likely action magnitude for the design criterion (e.g. the 'fifty-year wind' or the 'maximum credible earthquake').
4. Assess the structure in terms of its system properties.
5. Assess the influence function or functions operating between the action and the structural system (e.g. aerodynamic admittance).

6 Decide whether the approach will be spectrally based or time history based.
7 Combine the design action, influence function(s) and structural system properties to produce an estimate of the design response.
8 Remember to quote the response in terms of a probability of occurrence, unless the action is deterministic (i.e. it is definable with a simple mathematical description).
9 Proceed with a detailed design.
10 Take a look at the detailed design with the experiences contained in this book, and other 'experiential sources', in mind.

An explanation of these parts of the assessment is necessary.

Data about the action may be obtained from codes of practice, where they exist. These, however, should be treated with caution. Code committees often arrive at producing such documents through a process of compromises. The requirements should, therefore, be regarded by a designer as the minimum necessary to satisfy the requirements of the particular state or region with which he or she is dealing. If construction in a nearby region would involve the use of different design techniques, it is as well for the designer to be suspicious. There is the possibility that the code is out of date or has been produced through a less efficient process. In either case, erring on the safer side is recommended. It is also worth bearing in mind that simply increasing stiffness may not be the simplest or most efficient method of producing a safer design.

In the absence of guidance about actions likely in the region for the design, then the designer must use whatever information is available and use the techniques of Appendix E to try to arrive at a design loading. It may also be necessary to consider more than one action, and despite the fact that one may be a more stringent requirement than another, it may still be necessary to modify the design as a result. An example would be where both wind and earthquake actions may be encountered. If the wind action is larger, it may be tempting to discount the earthquake loading. However, this would be an error because the response of a structure to the two types of action is different. In particular, strength in the elastic range is the normal strategy to combat wind forces, whilst for earthquake action the provision of a ductile capability increases the chances of survival with cheaper maintenance costs.

Assessment of the structure in terms of its system properties should be facilitated by reference to previous chapters in this book. Nevertheless, a certain amount of intelligent interpretation may be necessary when moving from the style of cases presented here.

Influence functions such as wind parameters, or modal influence lines, can be found within this book. It is, however, absolutely essential that the designer tries to envisage the form of action that will occur, since this in turn may modify that influence. An example might be where the foundations of the structure are differentially placed in different types of soil when an earthquake action represents the design case.

A choice about whether to use a time history approach or a spectral approach is very much at the whim of the designer. If the data representing the loading form a random sequence, then it does, on the face of it, make sense to use a spectral approach. Such an approach is standard in wind engineering and earthquake engineering. However, if the modal response of the structure is likely to involve energy switching between modes of vibration, then a time history approach is strongly recommended. The reason for this is that it is normal that phase information is omitted when using a spectral representation of a forcing function. Since this is vital to an assessment of the relative response of different modes of vibration, then only through this medium is it possible to arrive at an estimate of the modal interaction that might occur.

When dealing with random loading functions, the probability of occurrence must be borne in mind. If the design parameters are couched in phrases that include the chance of the action occurring, then the untimely occurrence of the design event early in the life of the structure can be explained without embarrassment. Leaving the impression that the structure is safe for another 49 years after construction does no favours to the design team or to the profession.

8.7 *The use of models for design response prediction*

It is sometimes tempting to decide that the structure is complex, or that the loading will be complicated, so that building a model and testing it will be a simple method of arriving at a complete understanding of the likely structural dynamic response. Unfortunately, modelling for such actions as earthquakes and wind storms is itself a highly complex business, and interpretation is not straightforward. The fundamental reason is that, as soon as the scale is changed from 1:1, then some important parameters change at a different rate than do others. A simple example is that of the mass of a model. If the scale is chosen as being a factor of L when comparing model with prototype, then the mass will be a factor of L^3, because it is proportional to the length dimension times the breadth times the height, all of which are proportioned by L. As an example of the complexities, the following observations about wind-tunnel testing are offered.

The objective of making a model of a building or other structure and putting it in a wind tunnel is basically to try to see what the full-scale structure would do if it were subjected to extreme winds. Of course, it is not actually that simple. It is not even simple for those who know what they are doing. It is actually one of those branches of science that gets so close to being an art that it evokes some very strange responses from some of those who are close to the subject. Take, for instance, the statement from Professor Bob Scanlan who used to be at Princeton:

> Like in any good Fairy Story, the aerodynamicist has three wishes – and then he takes the consequences.

The three wishes are concerned with the rules of modelling and getting the model scales right. The three parameters are mass, length and time, and it is impossible to use the same scaling factor for all three at the same time except at a scale of 1:1. For instance, if we choose a scale of 10:1, then the lengths are scaled at a ratio of 10:1, but the mass is a three-dimensional quantity. The more height, breadth and width there is, the more the mass will increase. And so mass is proportional to length-cubed, and for a scale of 1:10, the mass will be 10^3 (or 1000) times smaller in the model. We may need to make up for this particular quirk by adding weights in strategic places. The upshot is that there are many conflicting requirements, and wind-tunnel testing is only valid if we resolve the conflicts.

We do this by considering that the scale of the model should be proportional to velocity-squared if we consider the gravity number similitude, that it should be inversely proportional to the velocity if we require Reynolds' number similitude, and that it should be proportional to velocity if there is similitude in the reduced frequency.

The art of modelling often includes the following considerations:

1 We do not need gravity number similitude unless gravity is significant in the motion of the structure.
2 Reynolds' number changes quite slowly with increasing velocity so long as there are no curved edges to the model. Often we can arrange that the flow is either subcritical or supercritical so long as it does not pass through the transition values.
3 We always need similitude in the reduced frequency since this effectively says that velocity equals length divided by time.

So it turns out that the reduced frequency is the only parameter that needs similitude between the model and the full scale, so long as you use a bit of fancy footwork to skip around the mathematical implications. The reduced frequency is

$$fD/V$$

where f is the natural frequency, D is an important dimension (diameter of a cylinder for instance) and V is the velocity of the wind. The value of this expression must be the same for the model and for the actual structure. So if we construct a model at a scale of 100:1, then D is going to have a ratio of 100 in it. The relationship for frequency is not always straightforward because it depends on the ratios of the stiffness of the model to the stiffness of the full-scale structure. But if we assume that the ratio is 10:1, then we are left with a requirement that the model-scale wind velocity will be scaled by a factor of 10 compared with the full scale. In other words, the wind velocity required in the wind tunnel will be 10 times slower than the design value for the full scale.

Once we are satisfied with the validity of the non-dimensional parameters (or at least that we can account for any difficulties), then we have to simulate the Earth's atmospheric boundary layer. Simply stated, the movement of air close to the ground experiences a slowing-down due to obstructions such as grass, trees, bushes, people, animals, houses, buildings, cities, hills and mountains. This means that the air closest to the ground moves most slowly (on average), and the higher you go the faster it gets, until the ground effect diminishes to nothing. The level at which this occurs is called the 'gradient height'. This happens at a height of about 1–2 km for normal winds, although for hurricanes and typhoons it may come down as low as 150 m, because the structure of those is completely different from complete frontal weather systems. Inside a wind tunnel, this boundary layer is 'grown' by placing roughness elements in the upstream section of the tunnel. At the working section, a check has to be made to ensure that the profile accurately models the profile to be expected at the precise location of the intended structure. To make sure this is correct, it is often necessary to include local topography such as surrounding buildings or hills. (Of course, if these surrounding buildings have sharp edges we run into problems with gravity number again.) The scale being used dictates whether we need local topography, local buildings or a combination of the two.

Having got the scaling right, we turn our attention to the measurement of the sway of the model. This can be done by using pressure taps inserted into the model. Each individual pressure tap responds to the pressure at the measurement position. This in turn is ducted through flexible tubing to a pressure-monitoring device. Very often there are many of these pressure taps and the amount of tubing is large. It is therefore led away to a 'scanivalve', which allows the measurement of pressures from one or a series of positions sequentially. The length and size of the tubing is carefully selected to avoid standing waves, which would lead to errors in the pressure indication.

An alternative to the measurement of pressure is a base bending moment measurement. The model is placed on a flexible mount and the small movements in three directions are measured. The measurement of the moment at the base of a rigid model can be transformed mathematically into the integrated force acting on the entire model. The characteristics of the structure itself, if there are dynamic motions to be included, can then be superimposed mathematically on top of the measured responses.

The suitably scaled model will be placed in a wind tunnel, and the correctly scaled wind will be allowed to blow at it. This wind will be subjected to roughness elements and the boundary layer will be 'grown',

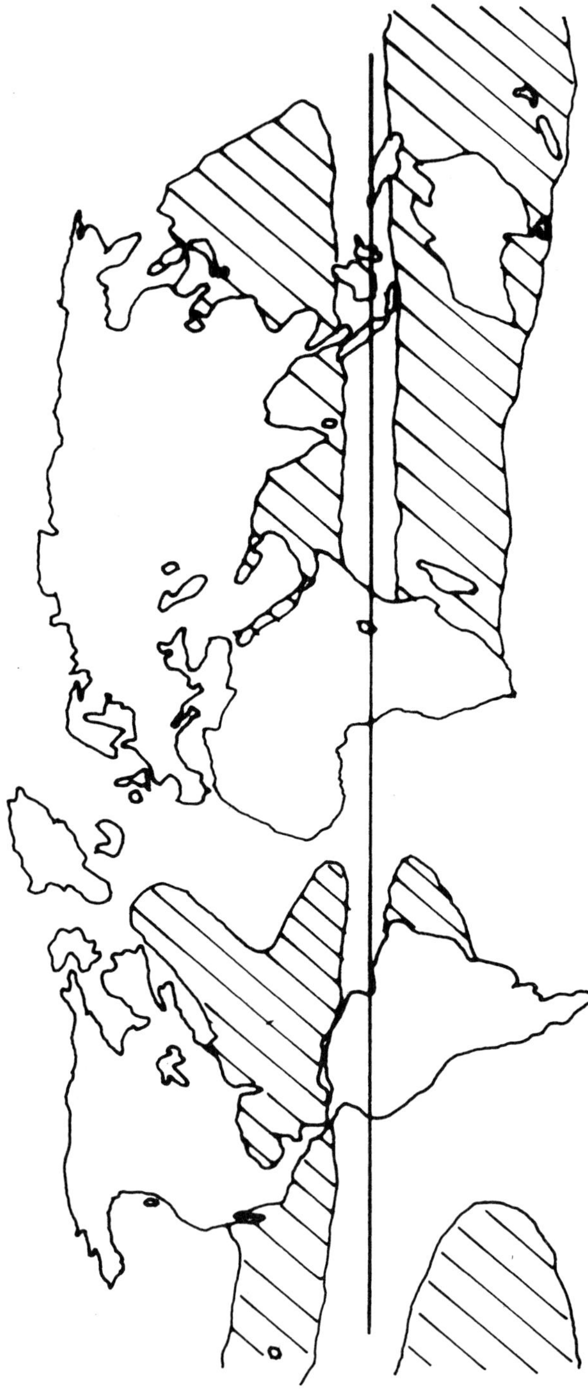

Fig. 8.3 Areas that would be affected by cyclones/typhoons for a 3°C warming of the sea surface.

until just close to the model under investigation the local topography will be simulated. This allows the estimation of the response for *one direction*. It may then be necessary to repeat the entire process for many directions, with appropriate modifications to the topography for each direction, of course.

After all of these processes, then we arrive at an estimate of what the response of the structure will be when subjected to the design wind. The accuracy of this estimate depends on the precision of the scaling and of measurement, the accuracy of the simulation, and the ability of those fairies to grant the three wishes.

8.8 *What if?*

Finally, the design should be assessed from the point of view of 'What if ...?'. The particular scenario that currently presents a potentially disastrous problem for a designer is associated with the warming Earth. The potential for tropical cyclones to travel outside their traditional territories is increasing. Indeed, such a move away from a long-term trend has been noted and shown to be statistically significant by the author. In this particular case, typhoons now show a greater tendency to recurve back into the Pacific when approaching the Eastern Asian coast. This, in turn, implies that the typhoon risk for Southern China has decreased, whilst at the same time it has increased for Japan, Korea and Northern China. From the same viewpoint, cities that are a little outside the traditional territories of tropical cyclones are likely to become at risk. This implies that an entirely new type of loading may occur. It is not simply a question of placing a multiplier on a current code of practice. It would be necessary to consider the pattern and magnitude of loading that occurs in an adjacent region. Figure 8.3 shows a representation of such regions were the sea surface temperature to increase, on average, by 3°C. Whilst no direct relationship between air temperature and sea surface temperature has been established, the Intergovernmental Panel on Climate Change (IPCC) has estimated that there is likely to be a 1–3°C rise in average global temperatures if we take no further action on reducing the so-called greenhouse gas emission intensity of today.

There are many uncertainties involved with the immediate future. It is probably wise to cater for such uncertainties, notwithstanding the requirements of formal legal frameworks. This is, of course, only achieved at the expense of some extra cost or of an intelligent modification of the available resources. It is hoped that this book gives some guidance on such efficiencies (the siting of columns and shear walls for instance), and that the informed designer can achieve the increased safety recommended without recourse to the need for increased spending.

APPENDIX A *Modified Mercalli scale*

I Not felt except by a very few under especially favourable circumstances.

II Felt only by a few persons at rest, especially on upper floors of buildings. Delicately suspended objects may swing.

III Felt quite noticeably indoors, especially on upper floors of buildings, but many people do not recognize it as an earthquake. Standing motor cars may rock slightly. Vibration like passing of truck. Duration estimated.

IV During the day, felt by many indoors, outdoors by few. At night, some awakened. Dishes, windows, doors disturbed; walls make creaking sound. Sensation like heavy truck striking building. Standing motor cars rocked noticeably.

V Felt by nearly everyone, many awakened. Some dishes, windows, etc., broken; a few instances of cracked plaster; unstable objects overturned. Disturbances of trees, poles and other tall objects sometimes noticed. Pendulum clocks may stop.

VI Felt by all, many frightened and run outdoors. Some heavy furniture moved; a few instances of fallen plaster or damaged chimneys. Damage slight.

VII Everybody runs outdoors. Damage negligible in buildings of good design and construction; slight to moderate in well-built ordinary structures; considerable in poorly built or badly designed structures; some chimneys broken. Noticed by persons driving motor cars.

VIII Damage slight in specially designed structures; considerable in ordinary substantial buildings, with partial collapse; great in poorly built structures. Panel walls thrown out of frame structures. Fall of chimneys, factory stacks, columns, monuments, walls. Heavy furniture overturned. Sand and mud ejected in small amounts. Changes in well water. Persons driving motor cars disturbed.

IX Damage considerable in specially designed structures; well designed frame structures thrown out of plumb; great in substantial buildings, with partial collapse. Buildings shifted off foundations. Ground cracked conspicuously. Underground pipes broken.

- X Some well-built wooden structures destroyed; most masonry and frame structures destroyed with foundations; ground badly cracked. Rails bent. Landslides considerable from river banks and steep slopes. Shifted sand and mud. Water splashed (slopped) over banks.
- XI Few, if any, (masonry) structures remain standing. Bridges destroyed. Broad fissures in ground. Underground pipelines completely out of service. Earth slumps and land slips in soft ground. Rails bent greatly.
- XII Damage total. Practically all works of construction are damaged greatly or destroyed. Waves seen on ground surface. Lines of site and level distorted. Objects are thrown upward into the air.

APPENDIX B Induced vibration testing on eight full-scale buildings

The results in Tables B.1–B.8 (taken from Jeary, 1986, pp. 741–2) relate to the first translational modes of eight BRE-tested buildings, some of which are mentioned in the text. Plans of the buildings are shown in Figs B.1–B.8.

Table B.1 South Stoneham House

Mode	*Force (N peak)*	*Freq. (Hz)*	*Amp. (mm)*	*Damping (%)*	*Modal mass* ($\times 10^6$*kg)*
NS1	7291	1.305	2.68	2.40	1.568
	5883	1.305	2.18	2.34	1.595
	4543	1.315	1.67	2.29	1.618
	3113	1.315	1.18	2.28	1.576
	1699	1.321	0.65	2.23	1.582
EW1	7573	1.330	2.66	2.75	1.379
	6156	1.335	2.23	2.56	1.425
	4718	1.340	1.71	2.44	1.483
	3233	1.340	1.26	2.48	1.357
	1761	1.345	0.68	2.38	1.417
	268	1.350	0.11	2.10	1.497

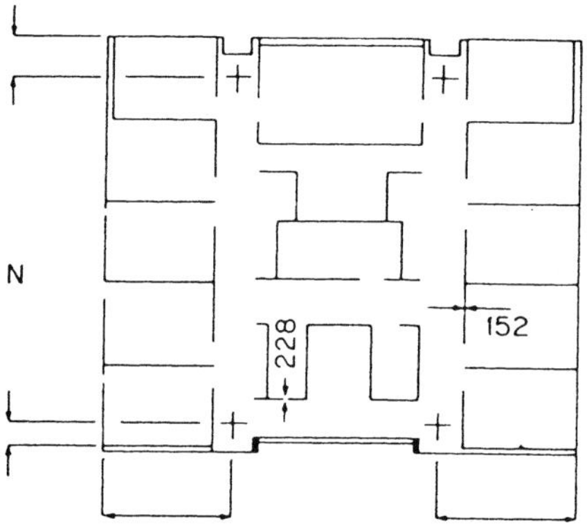

Fig. B.1 South Stoneham House.

Table B.2 Dunstan

Mode	*Force (N peak)*	*Freq. (Hz)*	*Amp. (mm)*	*Damping (%)*	*Modal mass (× 10^6kg)*
NS1	9000	1.45	2.75	2.60	0.875
	6850	1.46	2.09	2.43	0.924
	4650	1.47	1.42	2.42	0.914
	2350	1.48	0.75	2.29	0.916
	1450	1.48	0.47	2.18	0.938
	1000	1.49	0.33	2.11	0.942
	550	1.49	0.18	2.02	1.000
EW1	20900	2.21	1.73	3.24	1.209
	15850	2.22	1.32	3.16	1.218
	10900	2.23	0.90	3.00	1.259
	5400	2.25	0.46	2.94	1.236
	3350	2.26	0.29	2.85	1.264
	2300	2.27	0.20	2.69	1.320
	1250	2.28	0.11	2.57	1.287

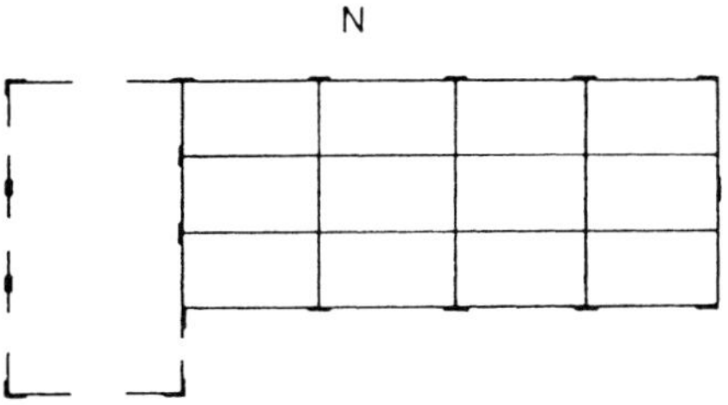

Fig. B.2 Dunstan.

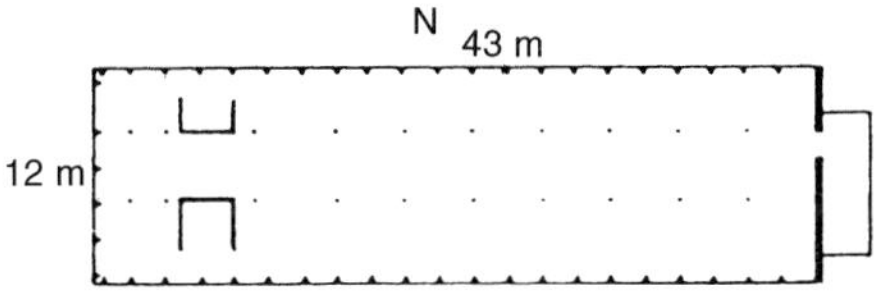

Fig. B.3 Sutherland House.

Table B.3 Sutherland House

Mode	*Force (N peak)*	*Freq. (Hz)*	*Amp. (mm)*	*Damping (%)*	*Modal mass (× 10^6kg)*
EW1	2800	1.46	0.41	2.77	2.930
	1945	1.47	0.29	2.44	3.222
	1110	1.51	0.16	2.11	3.653
	555	1.51	0.10	1.72	3.583
NS1	3931	1.73	0.72	1.49	3.102
	2725	1.74	0.48	1.32	3.598
	1500	1.755	0.28	1.32	3.337
	230	1.78	0.06	1.29	3.065

Table B.4 Leicester University Engineering Block

Mode	*Force (N peak)*	*Freq. (Hz)*	*Amp. (mm)*	*Damping (%)*	*Modal mass* ($\times 10^6 kg$)
NS1	1754	1.28	1.738	2.31	0.675
	1437	1.29	1.382	2.14	0.740
	1093	1.29	1.053	2.11	0.749
	761	1.30	0.745	2.22	0.689
	424	1.32	0.407	1.74	0.867
EW1	2376	1.49	1.240	2.74	0.798
	1943	1.50	0.998	2.71	0.810
	1498	1.51	0.730	2.54	0.897
	1054	1.53	0.496	2.54	0.905
	585	1.55	0.269	2.63	0.872

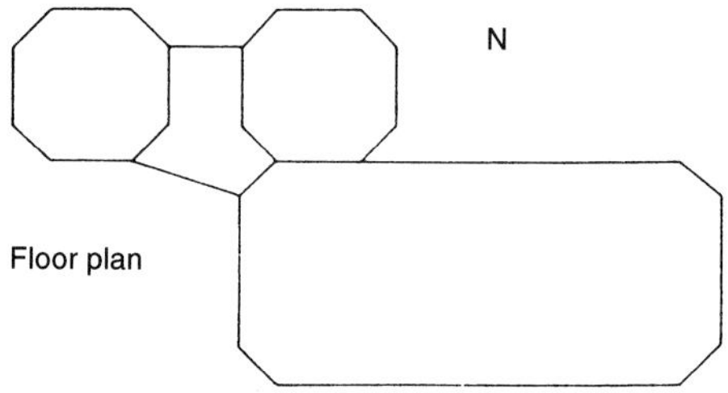

Fig. B.4 Leicester University Engineering Block.

Table B.5 John Russell Court building

Mode	*Force (N peak)*	*Freq. (Hz)*	*Amp. (mm)*	*Damping (%)*	*Modal mass* ($\times 10^6 kg$)
EW1	1954	0.972	1.81	1.15	2.52
	1589	0.977	1.47	1.00	2.85
	1191	0.980	0.84	0.99	3.80
	804	0.986	0.56	0.91	4.11
	402	0.986	0.43	0.84	2.88
	203	0.990	0.23	0.71	3.21
NS1	2778	1.159	1.402	1.44	2.59
	2245	1.165	1.162	1.21	2.98
	1687	1.166	0.870	1.24	2.91
	1131	1.169	0.577	1.27	2.86
	571	1.175	0.292	1.12	3.20
	289	1.182	0.176	1.08	2.76

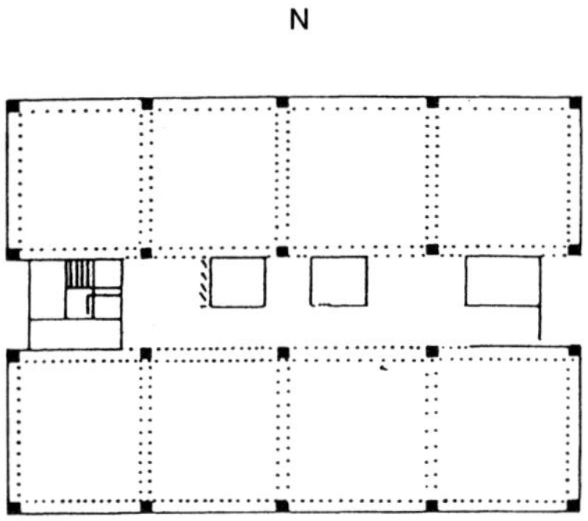

Fig. B.5 John Russell Court building.

Table B.6 Sheffield University Arts Tower

Mode	*Force (N peak)*	*Freq. (Hz)*	*Amp. (mm)*	*Damping (%)*	*Modal mass* ($\times 10^6$*kg)*
NS1	–	0.665	1.59	1.18	
	–	0.670	1.12	1.16	
	–	0.672	0.63	0.99	
	269	0.678	0.23	0.8	6.35
EW1	–	0.849	0.44	1.43	
	–	0.853	0.24	0.98	
	–	0.859	0.12	1.27	
	435	0.861	0.24	0.8	5.84

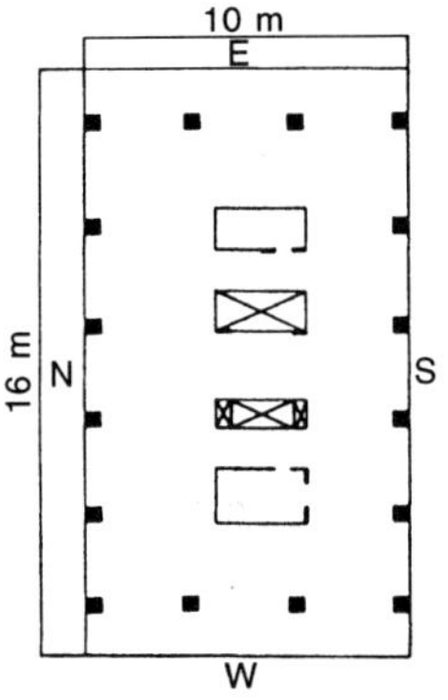

Fig. B.6 Sheffield University Arts Tower.

Table B.7 Priory Hall

Mode	*Force (N peak)*	*Freq. (Hz)*	*Amp. (mm)*	*Damping (%)*	*Modal mass* ($\times 10^6 kg$)
EW1	1463	0.841	1.52	1.39	2.57
	1182	0.845	1.21	1.47	2.59
	890	0.847	0.81	1.27	2.59
	599	0.851	0.65	1.43	2.56
	304	0.858	0.34	1.27	2.61
NS1	1683	0.902	1.38	1.96	2.12
	1352	0.904	1.13	1.83	2.13
	1026	0.909	0.89	1.71	2.11
	696	0.917	0.62	1.61	2.11
	353	0.923	0.34	1.44	2.16
	179	0.929	0.19	1.50	2.07

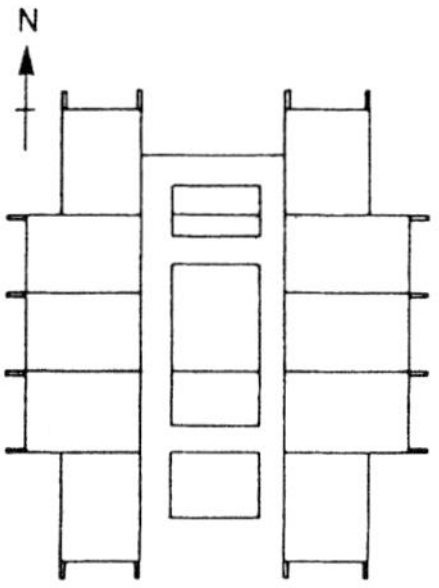

Fig. B.7 Priory Hall.

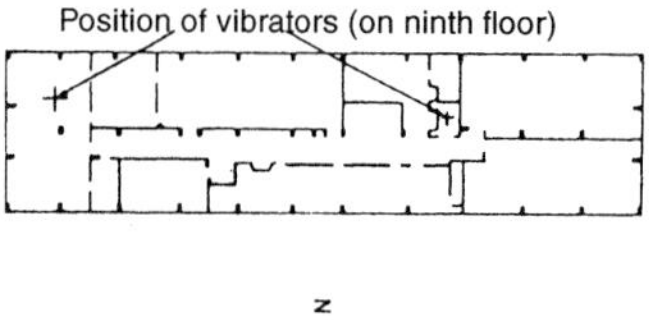

Fig. B.8 Exeter.

Table B.8 Exeter

Mode	*Force (N peak)*	*Freq. (Hz)*	*Amp. (mm)*	*Damping (%)*	*Modal mass* ($\times 10^6 kg$)
NS1	2087	1.393	0.432	2.3	4.05
	1674	1.392	0.357	2.3	3.96
	1273	1.392	0.250	2.3	4.32
	876	1.395	0.141	1.5	8.10
	475	1.396	0.046	1.5	13.4
	72	1.396	0.008	1.5	12.1
EW1	2904	1.647	0.512	3.4	6.18
	2346	1.648	0.417	3.2	6.48
	1788	1.650	0.325	1.6	9.60
	1226	1.650	0.162	1.4	15.0
	664	1.651	0.062	1.4	21.3
	100	1.651	0.012	1.4	16.8

APPENDIX C

Conventional analysis for statistical data

The conventional analysis in this appendix is carried out on real wind data from the city of Rio de Janeiro and its surrounding area. A map of the city and environs of Rio de Janeiro is shown in Fig. C.1. Wind data for Santa Cruz are listed in Table C.1, and results for three areas are plotted in Figs C.2–C.4.

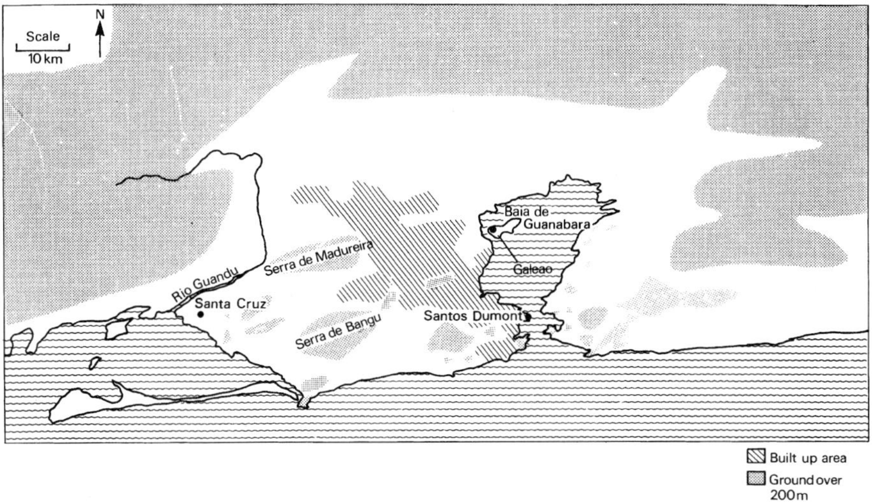

Fig. C.1 The city of Rio de Janeiro and its environs.

Table C.1 Santa Cruz

Date	*Independent storm values*					*Estimated annual values*	
	Max (knots)	*Run test*	*Dirn*	*Rank*	*Cumulative probability*	*Cumulative probability*	*Reduced variate*
02/01/71	25	+	240	27	0.443	0.008	1.527
07/03/71	35	+	90	5	0.820	0.000	2.709
24/04/71	28	+	210	25	0.410	0.005	1.677
27/05/71	30	+	210	12	0.197	0.000	2.278

Table C.1 (*Contd.*)

Date	*Independent storm values*					*Estimated annual values*	
	Max (knots)	*Run test*	*Dirn*	*Rank*	*Cumulative probability*	*Cumulative probability*	*Reduced variate*
22/06/71	30	+	20	13	0.213	0.000	2.227
12/08/71	22	−	360	55	0.902	0.537	−0.476
19/09/71	25	+	360	28	0.459	0.009	1.541
07/10/71	35	+	30	6	0.098	0.000	2.633
17/02/72	30	+	30	14	0.230	0.000	2.178
18/03/72	40	+	230	1	0.016	0.000	3.205
15/04/72	30	+	180	15	0.246	0.000	2.130
03/10/72	25	+	240	29	0.475	0.012	1.495
25/01/73	35	+	240	7	0.115	0.000	2.564
09/03/73	25	−	240	30	0.492	0.014	1.449
08/06/73	35	+	60	8	0.131	0.000	2.501
23/07/73	30	+	240	16	0.262	0.000	2.083
25/08/73	23	−	360	47	0.770	0.201	0.447
15/09/73	30	+	210	17	0.279	0.000	2.037
25/02/74	25	−	220	21	0.508	0.017	1.402
27/03/74	23	−	230	48	0.787	0.237	0.363
18/08/74	25	−	220	32	0.525	0.021	1.353
24/10/74	20	+	240	18	0.295	0.001	1.991
27/11/74	20	+	330	19	0.311	0.001	1.946
18/01/75	23	−	240	49	0.803	0.269	0.273
05/02/75	25	−	240	33	0.492	0.014	1.447
06/06/75	25	−	360	34	0.557	0.030	1.255
02/10/75	25	−	230	35	0.574	0.036	1.204
06/11/75	22	−	240	56	0.918	0.599	−0.667
14/12/75	22	−	220	57	0.934	0.666	−0.899
04/01/76	24	−	360	43	0.705	0.123	0.741
29/05/76	35	+	240	9	0.148	0.000	2.441
05/06/76	23	−	240	50	0.820	0.303	0.177
10/07/76	25	−	240	36	0.590	0.042	1.152
14/08/76	35	+	240	10	0.164	0.000	2.384
03/09/76	25	−	30	37	0.606	0.050	1.099
24/10/76	30	+	250	20	0.328	0.001	1.001
08/12/76	23	−	250	51	0.836	0.342	0.072
02/01/77	30	+	100	21	0.344	0.002	1.856
29/02/77	23	−	200	52	0.852	0.384	−0.043
01/04/77	25	−	250	38	0.623	0.058	1.044
16/05/77	30	+	250	22	0.361	0.002	1.811
24/07/77	40	+	240	2	0.033	0.000	3.021
14/08/77	38	+	180	4	0.066	0.000	2.794
14/09/77	30	+	260	23	0.377	0.003	1.767
10/11/77	25	−	360	39	0.639	0.068	0.987
05/12/77	32	+	240	11	0.180	0.000	2.330
02/02/78	23	−	360	53	0.791	0.245	0.341
29/03/78	22	−	220	58	0.951	0.739	−1.195
16/05/78	40	+	270	3	0.049	0.000	2.894
15/09/78	25	−	360	40	0.656	0.079	0.929

Table C.1 (*Contd.*)

Date	*Independent storm values*					*Estimated annual values*	
	Max (knots)	*Run test*	*Dirn*	*Rank*	*Cumulative probability*	*Cumulative probability*	*Reduced variate*
05/11/78	25	–	250	41	0.672	0.092	0.869
18/12/78	26	+	210	26	0.426	0.006	1.633
29/01/79	21	–	210	60	0.984	0.906	−2.311
24/05/79	25	–	200	42	0.689	0.107	0.806
16/07/79	23	–	210	54	0.885	0.481	−0.313
01/09/79	22	–	360	59	0.967	0.819	−1.609
15/10/79	24	–	210	44	0.721	0.141	0.673
16/12/79	24	–	230	45	0.738	0.161	0.602
17/08/80	30	+	230	24	0.393	0.004	1.722
30/10/80	24	–	330	46	0.754	0.184	0.527

$n = 60$, $r = 6.0$
Runs $R = 28$
$24 < R < 37$ for independence

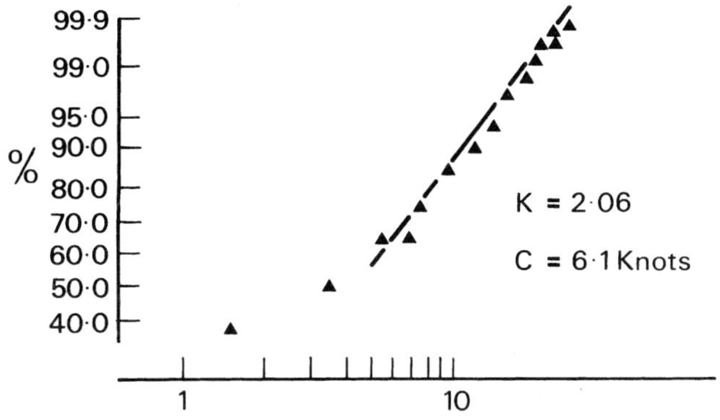

Fig. C.2 Parent distribution for Santa Cruz.

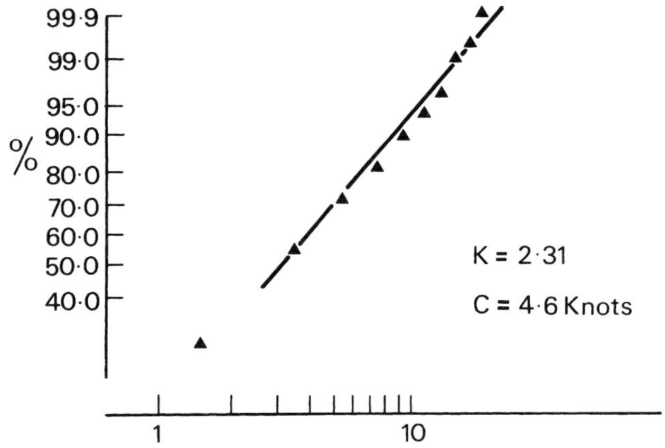

Fig. C.3 Parent distribution for Galeao.

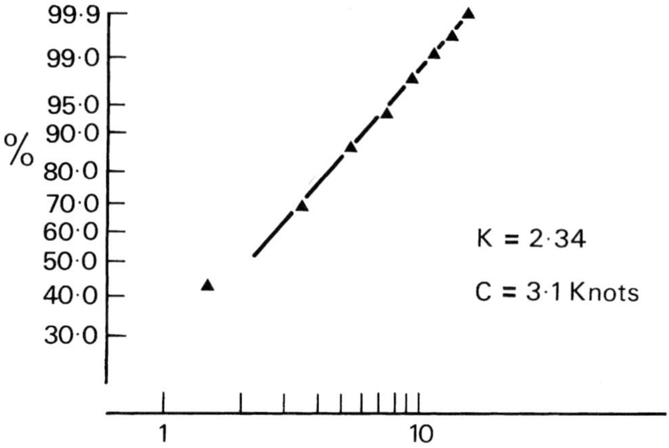

Fig. C.4 Parent distribution for Santos Dumont.

APPENDIX D

The use of the Leiblein BLUE estimator

In this appendix, we will be concerned with wind data from the city of Hong Kong, in particular Waglan Island. Table D.1 lists gust wind speeds (in knots) for Waglan Island from 1953 to 1980. The results are shown in Table D.2 and Figs D.1 and D.2.

Table D.1 Waglan gust wind speeds (knots) 1953–1980

n	$\hat{V}$	*Year*	*Rank*
1	93	1953	13
2	94	1954	11
3	94	1954	12
4	71	1957	21
5	100	1957	7
6	105	1960	5
7	69	1961	26
8	69	1961	27
9	117	1962	2
10	63	1963	33
11	76	1964	18
12	100	1964	8
13	124	1964	1
14	99	1964	9
15	71	1965	22
16	71	1966	23
17	62	1967	35
18	62	1967	36
19	52	1967	44
20	113	1968	3
21	50	1968	45
22	74	1969	20
23	83	1970	15
24	59	1970	38
25	55	1970	42
26	69	1971	28
27	80	1971	16
28	102	1971	6
29	86	1973	14
30	58	1974	39

Table D.1 (*Contd.*)

n	$\hat{V}$	*Year*	*Rank*
31	71	1974	24
32	55	1974	43
33	95	1975	10
34	69	1975	29
35	63	1976	34
36	70	1976	25
37	65	1977	32
38	76	1978	19
39	61	1978	37
40	69	1978	30
41	50	1978	46
42	56	1978	41
43	107	1979	4
44	80	1979	17
45	68	1980	31
46	57	1980	40

Maximum gusts ($\hat{V}$) in knots are noted for 46 sequential storms. These are ranked in order of gust velocity in the right-hand column. $X_n = N - \{2, 12, 30, 42\}$ in Table D.2 for the calculation of the BLUE coefficients.

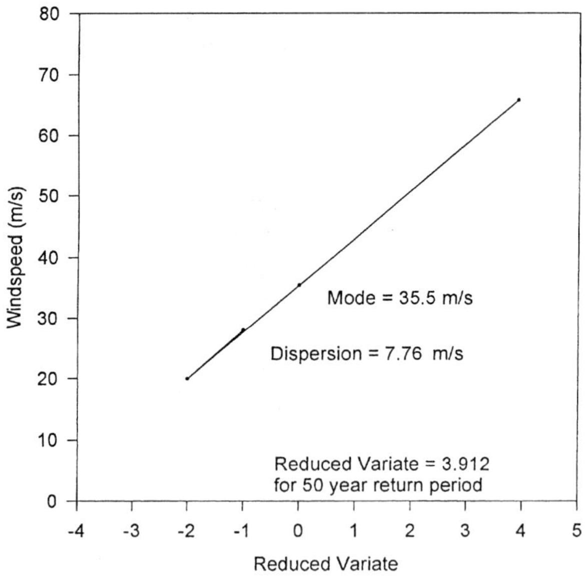

Fig. D.1 Maximum gust wind speed for Waglan Island (Hong Kong), fitted by Leiblein BLUE estimator.

Table D.2 BLUE estimator ($K = 4$) maximum Waglan gust (1953–1980)

X_n	$\hat{V}_{n_i}$	a_i^*	$a_i^* \hat{V}_{n_i}$	b_i^*	$b_i^* \hat{V}_{n_i}$
2	52	0.1893	9.8436	−0.3127	−16.2604
12	63	0.4566	28.7658	−0.1123	−7.0749
30	80	0.2772	22.176	0.2607	20.856
42	107	0.0768	8.2176	0.1643	17.5801
			69.003		15.1008

Therefore
mode = 69.003 knots $\equiv$ 35.4675 m/s
dispersion = 15.1008 knots $\equiv$ 7.7618 m/s
and so '50-year gust' is $35.4675 + 3.912 \times 7.7618 = 65.83$ m/s

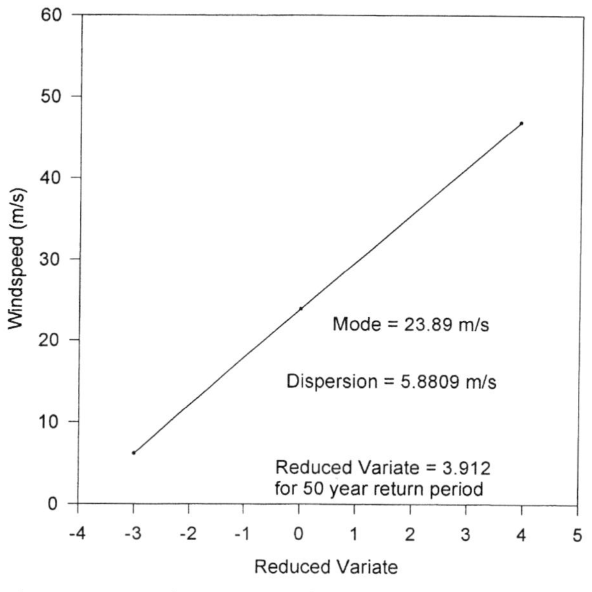

Fig. D.2 Annual mean wind speed for Waglan Island (Hong Kong), fitted by Leiblein BLUE estimator.

APPENDIX E

Conventional quasi-static approach and ESDU Item 76001

In this appendix, we look at the calculation of the response to wind of Priory Hall using the conventional quasi-static approach and the method of the Engineering Sciences Data Unit Item 76001. A plan of Priory Hall was given in Appendix B (Fig. B.7). Data relating to the building are presented in Table E.1. For the purposes of explanation, only the first EW mode of vibration is considered.

E.1 Quasi-static approach using the current HK code

The total force is

$$\begin{array}{rcrcrcr} 10 & \times & 21.8 & \times & 1.2 & = & 261.6 \\ 20 & \times & 21.8 & \times & 2.2 & = & 959.2 \\ 20 & \times & 21.8 & \times & 2.5 & = & 1090.0 \\ 6.4 & \times & 21.8 & \times & 3.0 & = & \underline{418.6} \\ & & & & & & 2729.4 \end{array} \tag{E.1}$$

or 2.7294 MPa. The quasi-static displacement for this building can be calculated as follows:

$$\text{modal stiffness } K = (2\pi f)^2$$

and

$$F/K = x$$

Thus

$$x = \frac{2.729 \times 10^6}{(2\pi \times 0.841)^2 \times 2.57 \times 10^6} = 0.038\ m \tag{E.2}$$

E.2 The ESDU 76001 method

The details of the calculation are contained in Table E.1. It can be seen that the peak response is $0.325g$. This can be converted to a displacement as follows:

$$x = \frac{0.325 \times 9.81}{4\pi^2 \times (0.841)^2} = 0.114\ m \tag{E.3}$$

Thus it can be seen that the complete response is composed of two parts (for the EW mode only) as follows: quasi-static=38 mm, dynamic=114 mm.

Table E.1 Priory Hall, Coventry: tabulated ESDU 76001 calculations for mean maximum acceleration

		EW *50-year*	*NS* *50-year*	*EW* *meas.*	*NS* *meas.*
Basic mean hourly wind speed at 10 m	$\bar{V}_{10}$	30	30	4.12(80)	4.12(80)
Surface roughness parameter	z_o	1	1	1	1
Power-law exponent	$\bar{\alpha}$	0.31	0.31	0.31	0.31
Displ. height of zero plane	d	7.5	7.5	7.5	7.5
Building height	H	56.4	56.4	56.4	56.4
Building width normal to wind dir.	W	21.8	18.8	21.8	18.8
Effective height	$\bar{z} = \frac{1}{2}(H - d)$	24.5	24.5	24.5	24.5
Mean hourly wind speed at z	$\bar{V}_z = \bar{V}_{10}(z/d)^{\bar{\alpha}}$	39.6	39.6	5.44	5.44
Mean hourly wind speed at H	$\bar{V}_H = \bar{V}_{10}(H/d)^{\bar{\alpha}}$	51.3	51.3	7.04	7.04
Resonance frequency	n	0.841	0.902	0.858	0.929
Damping	ζ	0.0134	0.0179	0.0118	0.0133
Modal mass	M	2.58E6	2.12E6	2.58E6	2.12E6
Turbulence length scale (x dir.)	${}^xL_u$	76.59	76.59	76.59	76.59
	$\tilde{n}_x = n\,{}^xL_u/\bar{V}_z$	1.62	1.74	12.08	13.08
Turbulence length scale (y dir.)	${}^yL_u$	33.72	33.72	33.72	33.72
	$\tilde{n}_y = n\,{}^yL_u/\bar{V}_z$	0.717	0.769	5.32	5.32
Turbulence length scale (z dir.)	${}^zL_u$	26.57	26.57	26.57	26.57
	$\tilde{n}_z = n\,{}^zL_u/\bar{V}_z$	0.564	0.605	4.19	4.54
	$nW/{}^yL_u$	0.544	0.503	0.547	0.518
	$nH/{}^zL_u$	1.785	1.915	1.821	1.972
Turbulence intensity	I_u	0.2767	0.2767	0.2767	0.2767
Spectral density function	ϕ_{uu}	0.0829	0.0791	0.0218	0.0207
	$\sqrt{A}/{}^zL_u$	1.320	1.226	1.320	1.226
Correction factor	B_n^2	1.92	1.95	1.93	1.97
Normalized force spectral function	J_H^2	0.118	0.118	0.118	0.118
Normalized force spectral function	J_W^2	0.21	0.21	0.21	0.21
Generalized modal force	$\bar{F}$	0.759E6	0.655E6	0.143E5	0.123E5
Resolved generalized modal force	$\bar{F}\cos\theta$	–	–	0.141E5	0.214E4
Spectral force function	$nS(n)$	0.696E9	0.502E9	0.635E5	0.142E4
Maximum dynamic response factor	g_D	4.148	4.165	4.153	4.172
Mean maximum acceleration	$\ddot{x}_{max}$	0.325	0.292	3.31E−3	5.69E−4

Mean hourly wind speed = 8 knots at 080° (max. gust 12); 8 knots = 8 × 0.5144 m/s = 4.12 m/s

APPENDIX F The Laplace transform

F.1 Definition and implications

The Laplace transform is defined in the following terms:

$$\mathscr{L} = \int_0^\infty e^{-st}\, dt \tag{F.1}$$

where s is a complex variable. This has several important implications. Firstly, the Laplace transform of a time function is

$$F(s) = \mathscr{L}[f(t)] \tag{F.2}$$

If these two equations are combined, we obtain

$$F(s) = \int_0^\infty f(t) e^{-st}\, dt \tag{F.3}$$

In other words, the Laplace transform of any time function can be obtained by integrating the product of the time history and e^{-st}.

We should also note the two very useful relationships:

$$\frac{1}{s} f(s) = \mathscr{L} \int_0^\infty f(t)\, dt \tag{F.4}$$

and

$$s f(s) = \mathscr{L} \frac{df(t)}{dt} \tag{F.5}$$

In words, these two equations state that a differential can be represented by multiplying by s, the Laplace operator, and an integral can be represented by dividing by s.

The approach that we can use is to represent ordinary differential equations with the Laplace operator replacing differentials. We may then rearrange the resulting algebraic equations and use look-up tables to find the inverse transform. This inverse function is a representation of a function of time. In other words, the time history of the response can be found from a look-up table.

F.2 Implementation

The equation for the response of a system represented in equation (3.8) can be replaced by its Laplace operator equivalent as follows:

$$MS^2 \cdot X(S) + CS \cdot X(S) + K \cdot X(S) = F(S) \tag{F.6}$$

we note the following relationships:

$$\zeta = \frac{C}{2\sqrt{KM}} \tag{F.7}$$

$$k = \omega^2 M \tag{F.8}$$

Thus we have the following:

$$MS^2 \cdot X(S) + 2\sqrt{KM} \cdot \zeta S \cdot X(S) + \omega^2 \cdot X(S) = F(S) \tag{F.9}$$

or

$$[S^2 + 2\zeta\omega S + \omega^2]X(S) = F(S)/M \tag{F.10}$$

This equation can be seen to be easily amenable to algebraic manipulation, and is a representation of a single-degree-of-freedom system. This can be represented in systems terms as

$$\frac{X(S)}{F(S)} = \frac{1/M}{S^2 + 2\zeta\omega S + \omega^2} \tag{F.11}$$

The right-hand side of the equation represents the 'transfer function'. We note that this is by definition written in S operator terms, and the relationship between response (X) and the input (F) in the Laplace plane is defined.

We can 'test' this system by subjecting it to various types of input. Examples might be a step function, a sine wave, or other more complex forms. Laplace transforms for the first two of these are given in Table F.1.

If we wish to see what the response of the system would be, we merely multiply the Laplace transform of the input by the Laplace transform of the single-degree-of-freedom system, and look up the result in a table of inverse transforms.

For instance, the response to a step function is

$$\frac{1}{S}\left(\frac{1/M}{S^2 + 2\zeta\omega S + \omega^2}\right) \tag{F.12}$$

or

$$\frac{1}{S}\left(\frac{1/M}{(S + \zeta w)^2 + \omega^2(1 - \zeta^2)}\right) \tag{F.13}$$

This may be found as relationship number 36 in the table of transforms by Day (1966):

- Laplace function

$$\frac{1}{S[(S+a)^2 b^2]} \tag{F.14}$$

- Time function

$$\frac{1}{a^2+b^2} + \frac{1}{b(a^2+b^2)^{1/2}} e^{-at} \sin(bt-\psi) \tag{F.15}$$

where $\psi = \tan^{-1}[b/(-a)]$. In this circumstance we note that $a = \zeta\omega$ and $b = \omega(1\text{-}\zeta^2)^{1/2}$, and by a straight substitution into the time function we obtain

$$X(t) = \frac{1}{\omega^2 M} + \frac{1}{\omega^2 M\sqrt{(1-\zeta^2)}} e^{-\zeta\omega t} \sin\left[\omega\sqrt{(1-\zeta^2)}\right] t - \psi \tag{F.16}$$

where $\psi = \cos^{-1}\zeta$ or

$$X(t) = \frac{1}{K}\left(1 - \frac{e^{-\zeta\omega t}}{\sqrt{(1-\zeta^2)}} \sin\left[\omega\sqrt{(1-\zeta^2)}\right] t - \psi\right) \tag{F.17}$$

This is the form of an exponential build-up of oscillation.

In the manner described above, the response of a single-degree-of-freedom system (or any other system for that matter) can be exposed to an input function of any form, and by algebraic manipulation the equations can be reduced to a form to be found in standard look-up tables.

Table F.2 lists some useful transform pairs. However, readers are referred to standard texts for fuller listings. Many more combinations are possible from this basic group by setting constants equal to zero.

Table F.1 Some Laplace transforms

Function	*Laplace transform*
unit step function	$1/S$
unit ramp	$1/S^2$
$\sin(\omega t)$	$\omega/(S^2 + \omega^2)$
$\cos(\omega t)$	$S/(S^2 + \omega^2)$
e^{-at}	$1/(S + a)$

Table F.2 Some useful transform pairs

Laplace transform	*Time function*
$\dfrac{1}{S+a}$	e^{-at}
$\dfrac{1}{S}$	$H(t)$ or 1
$\dfrac{1}{S^2}$	t
$\dfrac{1}{(S+a)^2+b^2}$	$\dfrac{1}{b}e^{-at}\sin(bt)$
$\dfrac{\omega}{(S+a)^2+\omega^2}$	$e^{-at}\sin(\omega t)$
$\dfrac{S}{S^2+a^2}$	$\cos(at)$
$\dfrac{S+b}{S^2+a^2}$	$\dfrac{1}{a(b^2+a^2)^{1/2}\sin(at+\Psi)}$ with $\Psi = \tan^{-1}(b/a)$
$\dfrac{S+c}{(S+a)^2+b^2}$	$\dfrac{1}{b}[(c-a)^2+b^2]^{1/2}\sin(bt+\Psi)$ with $\Psi = \tan^{-1}[b/(c-a)]$
$\dfrac{1}{S[(S+a)^2+b^2]}$	$\dfrac{1}{a^2+b^2}+\dfrac{1}{b(a^2+b^2)^{1/2}}e^{-at}\sin(bt-\Psi)$
$\dfrac{1}{(S+a)(S^2+b^2)}$	$\dfrac{e^{-at}}{a^2+b^2}+\dfrac{1}{b(a^2+b^2)^{1/2}}\sin(bt+\Psi)$ with $\Psi = \tan^{-1}(-b/a)$
$\dfrac{1}{(S+a)[(S+b)^2-c^2]}$	$\dfrac{e^{-at}}{(a-b)^2-c^2}+\dfrac{e^{-(b+c)t}}{2c(b+c-a)}+\dfrac{e^{(c-b)t}}{2c(a-b+c)}$

APPENDIX G

Calculation of added mass due to water behind a dam

An important parameter in the calculation of the response of dams is the effect of the added mass due to the existence of the water in the reservoir. The calculation of this added mass for the Emosson Dam is included in this appendix. In this particular case, the water level was 7 m below the level of the spillway.

G.1 Induced vibration – calculation of modal mass

The notation used in the following should be evident to the reader. The equation used has been presented in equation (3.28):

$$m_r = \frac{\hat{F}_r}{\hat{x}_r 8\zeta_r \pi^2 f_r^2} \tag{G.1}$$

For the first antisymmetric mode, this is

$$m_r = \frac{25.3 \times 10^3 \times 100}{21.2 \times 10^{-6} \times 8\pi^2 (1.85)^2 \times 2.1} = 2.103 \times 10^8 \text{ kg} \tag{G.2}$$

G.2 Calculation of the mass of the dam from drawings

We assume that the density of concrete is 2240 kg/m^3 and from the design the volume of the main part of the dam (excluding the wing dam) is 1.1×10^6 m^3. Then the mass of the dam is 24.64×10^8 kg, which can be expressed as

$$24.64 \times 10^8 = 0.658\gamma HLD \tag{G.3}$$

for the dam, with height H, width L, depth D and density of concrete γ.

G.3 Relationship between total mass and modal mass

It is assumed that the density of the dam is constant, and that the structure can be represented by an equivalent structure of constant width. A polynomial was fitted to half of the measured horizontal mode shape of the first mode. This mode was assumed to be antisymmetric. The mode and the polynomial fit are shown in Fig. 6.54.

The horizontal (upstream–downstream) displacement is denoted $\phi(h, y)$. Then the modal mass is given by

$$m_r = \int_0^D \int_0^H \int_0^L \phi^2(h, y)\, dy\, dh\, dp \tag{G.4}$$

At the crest the polynomial has the form

$$\begin{aligned}\phi(y) = &-0.04785 + 4.877(y/w) - 84.67(y/w)^2 + 610.6(y/w)^3 \\ &- 1512(y/w)^4 + 1184(y/w)^5\end{aligned} \tag{G.5}$$

for $0 < y < w/z$ at crest level. The variation of w with height was approximated as

$$w = bh/H \tag{G.6}$$

The variation of horizontal displacement with height was taken from measurements to be

$$\begin{aligned}\phi(h) &= 0 \qquad\qquad 0 < h < 0.28H \\ \phi(h) &= \frac{h - 0.28H}{0.72H} \qquad 0.28H < h < H\end{aligned} \tag{G.7}$$

The mode shape is therefore given by

$$\begin{aligned}\phi(h, y) = &\left(\frac{h - 0.28H}{0.72H}\right)\left[-0.04785 + 4.877\left(\frac{y}{w}\right) - 84.67\left(\frac{y}{w}\right)^2\right. \\ &\left. + 610.6\left(\frac{y}{w}\right)^3 - 1512\left(\frac{y}{w}\right)^4 + 1184\left(\frac{y}{w}\right)^5\right]\end{aligned} \tag{G.8}$$

The modal mass is calculated as

$$\begin{aligned}m_r &= 2\int_0^D \int_{0.28H}^H \int_0^{w/2} \gamma\phi^2(h, y)\, dy\, dh\, dp \\ &= 2\gamma D \int_{0.28H}^H \left(\frac{h - 0.28H}{0.72H}\right)^2 \int_0^{U/2} \left[-0.04785 + 4.877\left(\frac{y}{w}\right) - 84.67\left(\frac{y}{w}\right)^2\right. \\ &\quad\left. + 610.6\left(\frac{y}{w}\right)^3 - 1512\left(\frac{y}{w}\right)^4 + 1184\left(\frac{y}{w}\right)^5\right]^2 dy\, dh\end{aligned}$$

(constant thickness D assumed)

$$
\begin{aligned}
&= 2\gamma D \int_{0.28H}^{H} \left(\frac{h-0.28H}{H}\right)^2 \\
&\quad \times \left[0.0229y - \frac{0.466}{2}\left(\frac{y^2}{w}\right) + \frac{31.888}{3}\left(\frac{y^3}{w^2}\right) - \frac{884.305}{4}\left(\frac{y^4}{w^3}\right)\right. \\
&\quad + \frac{13269.499}{5}\left(\frac{y^5}{w^4}\right) - \frac{118260.361}{6}\left(\frac{y^6}{w^5}\right) \\
&\quad + \frac{640423.176}{7}\left(\frac{y^7}{w^6}\right) - \frac{2046952.960}{8}\left(\frac{y^8}{w^7}\right) \\
&\quad \left. + \frac{3732044.8}{9}\left(\frac{y^9}{w^8}\right) - \frac{3580416}{10}\left(\frac{y^{10}}{w^9}\right) + \frac{1401856}{11}\left(\frac{y^{11}}{w^{10}}\right)\right]_0^{w/2} \mathrm{d}h \\
&= 2\gamma D \int_{0.28H}^{H} \left(\frac{-0.28H}{0.72H}\right)^2 \times 0.1711 w \,\mathrm{d}h
\end{aligned}
\tag{G.9}
$$

Substituting equation (G.3) into (G.9) we have

$$
\begin{aligned}
m_{\mathrm{r}} &= 2\gamma D \int_{0.28H}^{H} \left(\frac{h-0.28H}{0.72H}\right)^2 \times 0.1711 \frac{hL}{H} \,\mathrm{d}h \\
&= 2\gamma D \int_{0.28H}^{H} 1.929\left(\frac{h^2}{H^2} - \frac{0.56H}{H^2}0.0784\right) \times 0.1711 \frac{hL}{H} \,\mathrm{d}h \\
&= 2 \times 1.929 \times 0.1711 \times \gamma D \left[\frac{h^4 L}{4H^3} - \frac{0.56h^3 HL}{3H^3} + \frac{0.0784h^2 L}{2H}\right]_{0.28H}^{H}
\end{aligned}
\tag{G.10}
$$

Therefore

$$
\begin{aligned}
m_{\mathrm{r}} &= 2 \times 1.929 \times 0.1711 \times 0.102 \times \gamma HLD \\
&= 0.06733\gamma HLD
\end{aligned}
\tag{G.11}
$$

This is equivalent to a modal mass of 2.41×10^8 kg. From section G.2 we have

$$
m_{\mathrm{r}} = 0.658\gamma HLD \tag{G.12}
$$

Using the value of modal mass calculated in section G.1, the total mass of the dam is calculated to be

$$
\frac{0.0658}{0.06733} \times 2.103 \times 10^8 \text{ kg} = 20.55 \times 10^8 \text{ kg} \tag{G.13}
$$

Bibliography

ANSI (1972) *American National Standard Building Code Requirements for Minimum Design Loads in Buildings and Other Structures*, ANSI A58.1, American National Standards Institute, New York.

Anthony, K.C. (1970) 'The background to the statistical approach', *The Modern Design of Wind Sensitive Structures, Proc. CIRIA Seminar*, London, pp. 17–27.

Arai, H., Hayashida, C. and Tsuchiya, J. (1962) *Study on Vertical Distributions of Wind Velocities and Evaluation of Angles of Sea Wind*, Railway Tech. Res. Rep. no. 315 (in Japanese).

Armer, G.S.T. (1977) *The Performance of Damaged Large Panel Structures*, BRE Seminar Paper, Building Research Establishment, Garston, England.

Armitt, J. and Counihan, J. (1968) 'The simulation of the atmospheric boundary layer in a wind tunnel', *Atmos. Environ.*, **2**, 49–71.

Barazangi, M. and Dorman, J. (1969) 'World seismicity maps compiled from ESSA, Coast and Geodetic Survey, epicenter data, 1961–67' *Bull. Seismol. Soc. Am.*, **59**, 369–380.

Bearman, P.W. (1971) 'Wind loads on structures in turbulent flow', *The Modern Design of Wind Sensitive Structures, Proc. CIRIA Seminar*, London.

Bendat, J. and Piersol, A. (1971) *Random Data: Analysis and Measurement Procedures*, Wiley, New York.

Bertero, V.V. (1973) 'Ductility and seismic response', *Planning and Design of Tall Buildings, Proc. 1972 ASCE–IABSE Int. Conf.*, Vol. 1b, ASCE, New York, pp. 303–309.

Blume, J. (1936) 'The building and ground vibrator', *US Coast and Geodetic Survey – Earthquake Investigations in California, 1934–1935*, Special Publ. no. 201.

Blume, J.A. (1960) 'Structural dynamics in earthquake-resistant design', *J. Struct. Div., ASCE*, **86**, no. ST4, Proc. Paper 1695, July 1958 with discussions ending September 1959; also in *Trans. ASCE*, **125**, 1088–1139.

Brownjohn, J.M.W. (1993) 'Damping in suspension bridges: sources, measurements and errors', *Third Asia–Pacific Symp. on Wind Engineering*, Hong Kong, December, pp. 217–222.

BSI (1972) *Code of Practice for the Structural Use of Concrete*, Part 1, *Design Materials and Workmanship*, CP110: Part 1, British Standards Institution, London (amended February 1976).

Cardenas, A.E. and Magura, D.D. (1973) 'Strength of high-rise shear walls – rectangular cross section', in *Response of Multistorey Concrete Structures to Lateral Forces*, SP-36, American Concrete Institute, Detroit, pp. 119–150.

Cermak, J.E. (1971b) *Wind Loading and Wind Effects*, *Proc. 5th Regional Conf.* (Chicago, IL), Lehigh University, Bethlehem, PA, pp. 49–52.

Chan, J.C.L. (1990) *Regional Interannual Variability of Tropical Cyclones Over the Western North Pacific*, CPHK Applied Science Res. Rep. AP-90-06.

Chen, T.Y. (1975) *Comparison of Surface Winds in Hong Kong*, Royal Observatory Tech. Rep. no. 41.

Chiu, A.N.L. and Nakamoto, R. (1981) *Experimental Study of Wind Effects on a Tall Guyed Tower: Report on Analysis of Wind Data*, Department of Civil Engineering, University of Hawaii at Manoa, Honolulu.

Chiu, A.N.L. and Nakamoto, R. (1985) 'Investigation of wind effects on a tall guyed tower', *J. Struct. Eng.*, **111**, no. 11.

Chiu, A.N.L. and Shi Yuan (1989) 'Full-scale study of wind effects on a tall structure', *Recent Advances in Wind Engineering*, *Proc. 2nd Asia–Pacific Symp. on Wind Engineering*, Vol. 2, International Academic Publishers, Beijing, pp. 697–700.

Choi, E.C.C. (1983a) 'Gradient height and velocity profile during typhoons', *6th Int. Conf. on Wind Engineering*, Australia.

Choi, E.C.C. (1983b) *Wind Loading in Hong Kong. Commentary on the Code of Practice on Wind Effects, Hong Kong – 1983*, Hong Kong Institution of Engineers.

Cole, H.A. Jr (1968) 'On-the-line analysis of random vibrations', *AIAA/ASME 9th Structural Dynamics Materials Conf.*, Palm Springs.

Cole, H.A. Jr (1973) *On-line Failure Detection and Damping Measurements by Random Decrement Signatures*, NASA-CR-2205.

Cook, N.J. (1973) 'On simulating the lower third of the urban adiabatic boundary layer in a wind tunnel', *Atmos. Environ.*, 7, 691–705.

Cook, N.J. and Prior, M.J. (1987) 'Extreme wind climate of the United Kingdom', *J. Wind Eng. Ind. Aerodyn.*, **26**, 371–389.

Crawford, R. and Ward, H.S. (1964) 'Determination of the natural periods of buildings', *Bull. Seismol. Soc. Am.*, **54**, no. 6.

Cummine, J., McCarthy, P. and Leduc, M. (1992) 'Blowdown over northern Ontario. A derecho event – July 18, 1991', *4th AES/CMOS Workshop on Operational Meteorology*, Whistler, British Columbia, Atmospheric Environment Services. Canada, Meteorological and Oceanographic Society, pp. 311–317.

Dampier, W. (1697) *A Voyage Round the World.*

Davenport, A.G. (1961a) 'The spectrum of horizontal gustiness near the ground in high winds', *Q. J. R. Meteorol. Soc.*, **87**, 194–211.

Davenport, A.G. (1961b) 'The application of statistical concepts to the wind loading of structures', *Proc. Civ. Eng.*, **19**, 449.

Davenport, A.G., (1965) 'The relationship of wind structure to wind loading', *Proc. Conf. on Wind Effects on Buildings and Structures*, National Physical Laboratory, 1963, HMSO, London.

Davenport, A.G. (1989) *Proposed New International (ISO) Wind Load Standard*, P1199–1214, APSOWE 2.

Davenport, A.G. and Hill-Carroll, P. (1986) 'Damping in tall buildings: its variability and treatment in design', *ASCE Conf.*, Seattle.

Davenport, A.G., Georgiou, P.N., Mikitiuk, M., Surry, D. and Lythe, G. (1984) 'The wind climate of Hong Kong', *3rd Int. Conf. on Tall Buildings*, Hong Kong and Guangzhou, pp. 454–460.

Davenport, A.G., Isyumov, N., Lythe, G.R., Robertson, L.E., Steckley, A.C. and Surry, D. (1988) 'The wind engineering study for the Bank of China', *Proc. 4th Int. Conf. on Tall Buildings*, Hong Kong University Press, pp. 143–147.

Deaves, D.M. and Harris, R.I. (1978) *A Mathematical Model of the Structure of Strong Winds*, CIRIA Rep. 76.

Deinum, P.J., Dungar, R., Ellis, B.R., Jeary, A.P., Severn, R.T. and Read, G.A.L. (1982) 'Vibration tests on Emosson Arch Dam, Switzerland', *Earthq. Eng. Struct. Dyn.*, **10**, 447–470.

Desouza, V.C.M., Jeary, A.P. and Sparks, P.R. (1980) 'Concideracoes sobre o comportamento dinamico de edificios premoldados', *Rev. Brasil. Technol.*, pp. 219–231.

Dharan, L.R. and Recker, R.L. (1991) 'Growth of longitudinal matrix damage in unidirectional composites', *Eng. Fract. Mech.*, **38**, no. 2/3, 185–195.

Duchene-Marullaz, P. (1975) 'Full-scale measurements of atmospheric turbulence in a suburban area', *Proc. Fourth Int. Conf. on Wind Effects on Buildings and Structures*, pp. 23–32.

Eaton, K.J. (1975) *Cladding and the Wind*, BRE Current Paper 47–75, HMSO, London.

Eaton, K.J. and Mayne, J.R. (1975) 'The measurement of wind pressures on two-storey houses at Aylesbury', *J. Ind. Aerodyn.*, **1**, no.1, 67–109.

Ellis, B.R. (1979) 'A study of dynamic soil–structure interaction', *Proc. Inst. Civ. Eng., Part 2*, **67**, 771–783.

Ellis, B.R. (1980) 'An assessment of the accuracy of predicting the fundamental natural frequencies of buildings and the implications concerning dynamic analysis of structures', *Proc. Inst. Civ. Eng., Part 2*, **69**, 763–776.

Ellis, B.R. and Jeary, A.P. (1980a) 'A review of recent work on the dynamic behaviour of tall buildings at various amplitudes', *6th World Conf. on Earthquake Engineering*, Istanbul.

Ellis, B.R. and Jeary, A.P. (1980b) *Vibration Testing of Fixed Offshore Structures*, Report of Offshore Technology Board OT-R-8111.

Ellis, B.R. and Jeary, A.P. (1984) 'On the forced vibration testing of dams', *7th World Conf. on Earthquake Engineering*, San Francisco.

Ellis, B.R. and Littler, J.D., (1987) 'Dynamic response of nine similar tower blocks', *Seventh Int. Conf. on Wind Engineering*, Aachen, FDR.

Ellis, B.R., Jeary, A.P. and Severn, R.T. (1984) 'Integrity monitoring of dams by dynamic tests', *Proc. Int. Conf. on Safety of Dams*, Coimbra, Portugal, pp. 147–152.

ESDU (1982) *Strong winds in the atmospheric boundary layer: Part 1 Mean hourly wind speeds; Part 2 Discrete gust speeds*, Engineering Sciences Data Item 82026, ESDU International, London.

Esteva, L. (1970) 'Practical considerations in the Bayesian estimate of seismic risk' (Consideraciones practicas en la estimacion Bayesiana de riesgo sismico), *Ingenieria*, **40**, no. 2, 209–228.

Esteva, L. (1973) 'Summary report: structural safety and probabilistic methods', *Planning and Design of Tall Buildings* (*Proc. 1972 ASCE–IABSE Int. Conf.*), Vol. 1b, ASCE, New York, pp. 1043–1066.

Esteves, J.M. (1978) *Control of Vibration Caused by Blasting*, Laboratorio Nacional de Engenharia Civil, Lisbon, Memoria 498.

Eyre, R. and Tilley, G.P. (1977) 'Damping measurements on steel and composite bridges', *Symp. on Dynamic Behaviour of Bridges*, Transport and Road Research Laboratory, Crowthorne, Berkshire, UK.

Flay, R.G.J. and Stevenson, D.C. (1988) 'Integral length scales in strong winds below 20 m', *J. Wind Eng. Ind. Aerodyn.*, **28**, 21–30.

Foutch, D.A. (1978) 'The vibrational characteristics of a twelve storey steel frame building', *Earthq. Eng. Struct. Dyn.*, **6**, 265–294.

Fujino, Y., Pachero, B.M., Sun, L.M. and Chaiseri, P. (1989) 'Tuned liquid damper for suppressing horizontal motion of structures', *Int. Workshop*, *'Damping 89'*, Florida.

Galambos, T.V. and Mayes, R.C. (1979) 'Lessons from dynamic tests of an eleven storey building', *Eng. Struct.*, **1**, October.

Gao, X.F., Chung, C.Y., Jeary, A.P., Wong, J.C.K. and An, Y. (1997) 'The effect of typhoons on the microstructure of concrete in a reinforced structure', *J. Cement Concrete Compos.*, (not yet assigned).

Georgiou, P.N., Surry, D. and Davenport, A.G. (1988) 'Codification of wind loading in a region with typhoons and hills', *Fourth Int. Conf. on Tall Buildings*, Hong Kong, pp. 252–258.

Griffith, A.A. (1920) 'Theory of rupture of brittle materials', *Trans. R. Soc., Ser. A*, **21**, 163–198.

Harris, R.I. (1968) *On the Spectrum and Autocorrelation Function of Gustiness in High Winds*, ERA Rep. 5273.

Harris, R.I. (1971) *The Nature of the Wind*, Paper 3, *The Modern Design of Wind-Sensitive Structures*, CIRIA, pp. 29–56.

Hart, G.C. (1969) 'Combining a wind tunnel analysis with a three-dimensional analytic building analysis', *Proc. Technical Meeting Concerning Wind Loads on Buildings and Structures*, NBS, Gaithersburg, MD, Building Science Series 30.

Hart, G.C. and Vasudevan, R. (1975) 'Earthquake design of buildings: damping', *J. Struct. Div., ASCE*, Issue ST1, 11–30.

Hart, G.C., DiJulio, R.M. Jr and Lew, M. (1974) 'High-rise building response: damping and period nonlinearities', *Proc. 5th World Conf. on Earthquake Engineering*, Rome.

Hassett, T.C., Jeary, A.P., Mo, P.T. and Lau, K.C. (1987) 'A centralised laboratory system in a tertiary institution', *Hong Kong Engineer, Proc. HKIE*, August.

Hill-Carroll, P.E.B. (1985) 'The prediction of mean structural damping values and their coefficient of variation', Master of Engineering Science Thesis, University of Western Ontario, London, Canada.

Hisada, T. (1971) 'Earthquake resistant design of high-rise buildings in Japan', *3rd Regional Conf. Proc.* Muto Institute of Structural Mechanics, Tokyo, pp. 90–98.

Hisada, T. (1973) 'Loading and response criteria and code requirements', *Planning and Design of Tall Buildings, Proc. 1972 ASCE–IABSE Int. Conf.*, Vol.1b, ASCE, New York, pp. 267–278.

Hisada, T. and Nakagawa, K. (1956) 'Vibration of buildings in Japan', *Proc. 1st World Conf. on Earthquake Engineering*, Berkeley, CA.

Holford, I. (1977) *The Guinness Book of Weather Facts and Feats*, Guinness Superlatives, Enfield, UK.

Houghton, J.T., Jenkins, G.J. and Ephraums, I.J. (1990) *Climate Change – The IPCC Scientific Assessment*, Cambridge University Press.

Housner, G.W. (1947) 'Characteristics of strong motion earthquakes', *Bull. Seismol. Soc. Am.*, **31**, no. 1.

Housner, G.W. (1956) 'Limit design of structures to resist earthquakes', *Proc. 1st World Conf. on Earthquake Engineering*, Berkeley, CA.

Housner, G.W. (1973) 'Earthquake ground motion', *Planning and Design of Tall Buildings, Proc. 1972 ASCE–IABSE Int. Conf.*, Vol.1b, ASCE, New York, pp. 159–176.

Hudson, D.E. (1964) 'Resonance testing of full-scale structures', *J. Eng. Mech. Div., ASCE*, **90**, no. EM3, Paper 3923.

Hudson, D.E. (1971) *Strong Motion Instrumental Data on the San Fernando Earthquake of February 9, 1971*, California Institute of Technology, Pasadena, CA.

Hudson, D.E., Alford, J.L. and Housner, G.W. (1954) 'Measured response of a structure to an explosive-generated ground shock', *Bull. Seismol. Soc. Am.*, **44**, no.3.

Hunt, J.C.R. and Durbin, P. (1979) 'A theory for fluctuating surface pressures on structures in turbulent winds', *Proc. Fifth Int. Conf. on Wind Effects on Buildings and Structures*, Fort Collins.

Ishizaki, H. (1971) 'Storm frequencies and wind load problems', *Proc. Third Int. Conf. on Wind Effects on Buildings and Structures*, Tokyo, pp. 539–548.

Ishizaki, H. (1983) 'Wind profiles, turbulence intensity and gust factors for design in typhoon-prone regions', *6th Int. Conf. on Wind Engineering*, Australia.

Jacobsen, L.S. (1965) 'Damping in composite structures', *Proc. 2nd World Conf. on Earthquake Engineering*, pp. 1029–1044.

Jeary, A.P. (1974) 'Damping measurements from the behaviour of several large multi-flue chimneys', *Inst. Civ. Eng. Proc.*, Paper 7967, June.

Jeary, A.P. (1983) *Checking the Probable Response of Suspended Floors*, BRE Information Paper 17/83.

Jeary, A.P. (1986a) 'Damping in tall buildings', *Conf. on the Second Century of the Skyscraper*, ASCE/Council on Tall Buildings, Chicago.

Jeary, A.P. (1986b) 'Damping in tall buildings. A mechanism and a predictor', *Int. J. Earthq. Eng. Struct. Dyn.*, **14**, 733–750.

Jeary, A.P. (1986c) 'In-service performance of bridges', *China Civil Engineering Society Congress*, Shanghai.

Jeary, A.P. (1987a) 'Determination of the micro-cracking regime in tall buildings', *7th Int. Wind Conf.*, Aachen.

Jeary, A.P. (1987b) 'The vibration response of suspended floors to crowd loading', *Conf. Internoise 1987*, Beijing, pp. 567–570.

Jeary, A.P. (1990) 'The impact of modulated feedback on design criteria in a non-linear environment', *Tall Buildings 2000 and Beyond*, Council on Tall Buildings and Urban Habitat, pp. 693–712.

Jeary, A.P. (1991) 'A commentary on future codes of practice', *8th Int. Conf. on Wind Engineering*; and *J. Wind Eng. Ind. Aerodyn.*, **41–44** (1992), 2895–2906.

Jeary, A.P. (1992) 'Establishing non-linear damping characteristics of structures from non-stationary response time-histories', *Struct. Eng.*, **70**, no. 4, 61–66.

Jeary, A.P. (1995a) 'Integrity monitoring of structures', *9th Int. Conf. on Wind Engineering*, New Delhi; and *Wind Engineering Retrospect and Prospect*, Vol. 4, pp. 2189–2200.

Jeary, A.P. (1995b) 'The description and measurement of non-linear damping in structures', *IWEF Conf. on Damping in Structures*, Atsugi City, Japan.

Jeary, A.P. and Ellis, B.R. (1981a) 'Vibration tests on structures at various amplitudes', *ASCE/EMD Conf. on Dynamics of Structures*, Atlanta, GA, pp. 281–294.

Jeary, A.P. and Ellis, B.R. (1981b) 'The accuracy of mathematical models of structural dynamics', *Inst. Struct. Engs./BRE Int. Seminar on Dynamic Modelling of Structures*, Garston, England.

Jeary, A.P. and Ellis, B.R. (1984) 'Non-destructive in-situ testing using dynamics techniques', *Third Int. Conf. on Tall Buildings*, Hong Kong and Guangzhou, Paper T70, pp. 76–81.

Jeary, A.P. and Irwin, A.W. (1976) 'Vibrations of a nuclear power station charge hall', *Int. J. Earthq. Eng. Struct. Dyn.*, **4**, 221–229.

Jeary, A.P. and Sparks, P.R. (1977) 'Some observations on the dynamic sway characteristics of concrete structures', *Symp. on Vibrations in Concrete Structures*, New Orleans, and BRE Current Paper CP 7/78.

Jeary, A.P. and Winney, P.E. (1972) 'Determination of structural damping of a large multi-flue chimney from the response to wind excitation', *Proc. Inst. Civ. Eng., Part 2*, **53**, 569–577.

Jeary, A.P., Ellis, B.R. and Vasconcellos, R. (1979) 'The practical vibration analysis of structures with particular reference to offshore structures', *Conf. Brasil Offshore*, Session 6 COPPE, Universidad Federal de Rio de Janeiro.

Jeary, A.P., Sparks, P.R. and Desouza, V.C.M. (1981) 'A study of the use of ambient vibration measurements to detect changes in the structural characteristics of a building', *ASCE/EMD Conf. on Dynamics of Structures*, Atlanta, GA, pp. 189–199.

Jeary, A.P., De Souza, V.C.M. and Vasconcellos, R. (1984a) 'Inspection of a railway bridge by dynamic test', *Conf. INERC, Comportarea in situ a Constructilor*, Bucharest, pp. 229–238.

Jeary, A.P., Beak, M., Ellis, B.R. and Littler, J.D. (1984b) *Report on Vibration Tests at Ronan Point*, BRE A200/84/59.

Jeary, A.P., Morris, R.G. and Tomlinson, R.W. (1988) 'Human perception of motion. Tests in tall buildings', *7th Int. Wind Conf.*, Aachen; and *J. Wind. Eng. Ind. Aerodyn.*, **29**, 361–370.

Jennings, P.C., Matthiesen, R.B. and Hoerner, J. (1972) 'Forced vibration test of a twenty two storey steel frame building', *J. Earthq. Eng. Struct. Dyn.*, **1**, 107–132.

Jones, P.D. (1989) *United Nations Environmental Data Report*, Blackwell, Cambridge, MA and Oxford.

Kareem, A. (1986) 'Performance of cladding in hurricane Alicia', *Proc. ASCE. J. Struct. Eng.*, **112**, 2679–2693.

Kato, N., Ohkuma, J.R., Kim, J.R., Marukawa, H. and Niihori, Y. (1992) 'Full scale measurements of wind velocity in two urban areas using an ultrasonic anemometer', *8th Int. Conf. on Wind Engineering*; and *J. Wind Eng. Ind. Aerodyn.*, **41–44**, 67–78.

Koen, B.V. (1984) 'Towards a definition of the engineering method', *Eng. Educ.*, December, pp. 150–5.

Kurdowski, W. and Sorrentino, F. (1983) 'Special cements', in *Structure and Performance of Cements*, ed. P. Barnes, Applied Science, Barking, UK.

Lagomarsino, S. (1993) 'Forecast models for damping and vibration periods of buildings', *J. Wind Eng. Ind. Aerodyn.*, **48**, 221–239.

Laneville, A. and Parkinson, G.V. (1971) 'Effects of turbulence on galloping of bluff cylinders', *Proc. 3rd Int. Conf. on Wind Effects on Buildings and Structures*, Tokyo.

Lee, B.E. (1987) 'Engineering design for extreme wind occurrence in Hong Kong', *US–Asia Conf. on Engineering for Mitigating Natural Hazards Damage*, Bangkok, Paper A8.

Littler, J.D. and Ellis, B.R. (1987) 'Vibration studies on the Humber bridge', *Conf. on Flow Induced Vibrations*, Bowness-on-Windermere, UK.

Littler, J.D. *et al.* (1988) 'The measurement of the dynamic response of Sheffield Arts Tower to wind loading', *Int. Conf. on Tall Buildings*, Hong Kong and Shanghai.

Loland, O. and Dodds, C.J. (1976) 'Experiences in developing and operating integrity monitoring systems in the North Sea', *Offshore Technology Conf.*, Aberdeen, Paper 2551.

Mackey, S. and Ko, P.K.L. (1975) 'Spatial configuration of gusts', *Proc. Fourth Int. Conf. on Wind Effects on Buildings and Structures*, pp. 41–52.

Maeda, J. and Makino, M. (1988) 'Power spectra of longitudinal and lateral wind speed near the ground in strong winds', *J. Wind Eng. Ind. Aerodyn.*, **28**, 31–40.

Mahmoodi, P., Robertson, L.E., Yontar, M., Moy, C. and Feld, L. (1989) *Performance of Viscoelastic Structural Dampers in World Trade Center Towers*, Report of 3M Co., St Paul, Minnesota.

Maribo Pedersen, B., Hansen, K.H., Oye, S., Brinch, M. and Fabian, O. (1992) 'Some experimental investigations on the influence of the mounting arrangements on the accuracy of cup-anemometer measurements', *J. Wind Eng. Ind. Aerodyn.*, **39**, 373–383.

McNichol, D.P. (1990) 'Public housing blocks in Hong Kong: the identification, investigation and rectification of structural defects', *Struct. Eng.*, **68**, no. 16, 307–316.

Meek, C.E., Reid, I.M. and Manson, A.H. (1985) 'Observation of mesosophic wind velocities', *Radio Sci.*, **20**, no. 6, 1383–1402.

Meister, R.I. (1937) 'The physiological evaluation of vibration measurement' (Die Physiologische Wertung von Erschutterungsmessungen), *Akustische Z.*, **2**, 1–10.

Melbourne, W.H. (1972) 'Modelling of structures to measure wind effects', *Proc. Conf. on Structural Models*, Sydney, Australia, Cement and Concrete Association of Australia in conjunction with Department of Architectural Science, University of Sydney, and Institution of Engineers, Australia, NSW Division.

Melbourne, W.H. (1974) 'Peak factors for structures oscillating under wind action', *Proc. Conf. on the Application of Probability Theory to Structural Design*, Institution of Engineers, Australia, pp. 34–44.

Melbourne, W.H. (1975) 'The relevance of codification to design', *Proc. 4th Int. Conf. on Wind Effects on Buildings and Structures*, London, University of Cambridge Press.

Melbourne, W.H. (1984) 'Design wind data for Hong Kong and surrounding countryside', *3rd Int. Conf. on Tall Buildings*, Hong Kong and Guangzhou, pp. 461–467.

Melbourne, W.H. and Joubert, P.N. (1971) 'Problems of wind flow at the base of tall buildings', *Proc. Int. Conf. on Wind Effects on Buildings and Structures*, Tokyo.

Mickle, R.E., Cook, N.J., Hoff, A.M., Jensen, N.O., Salmon, J.R., Taylor, P.A., Tetzloff, G. and Teunissen, H.W. (1988) 'The Askervein Hill Project: vertical profiles of wind and turbulence', *Boundary Layer Meteorol.*, **43**, 143–169.

Mindess, S. and Young, J.F. (1981) *Concrete*, Prentice-Hall, Englewood Cliffs, NJ.

Miyashita, K., Tamura, Y., Asami, Y. and Naito, S. (1993) *Third Asia–Pacific Symp. on Wind Engineering*, Hong Kong, pp. 423–428.

Morris, D. (1978) *Manwatching. A Field Guide to Human Behaviour*, Triad/Panther.

Naito, G. (1988) 'Turbulent properties and spectral behaviours of ocean winds observed at an offshore tower', *J. Wind Eng. Ind. Aerodyn.*, **28**, 51–60.

Natrella, M.G. (1966) *Experimental Statistics*, National Bureau of Standards.

Newberry, C.W. and Eaton, K.J. (1974) *Wind Loading Handbook*, BRE Report, HMSO, London.

Ostrowski, J., Jeary, A.P. and Grace, W.R. (1989) 'The role of vibration testing for assessment of a damaged wharf', *Int. Conf. on Case Histories in Structural Failures*, Singapore.

Peterka, J.A. and Shahid, S. (1993) 'Extreme gust winds in the USA', *7th US National Conf. on Wind Engineering*, UCLA, pp. 503–512.

Powell, M.D. (1979) 'An evaluation of diagnostic marine boundary layer models applied to tropical cyclones', *5th Int. Conf. on Wind Engineering*, Colorado, pp. 133–144.

Reynolds, R. (1988) *J. Climate*, no. 1, 75–86.

Richter, C.F. (1958) *Elementary Seismology*, W.H. Freeman, San Francisco.

Robertson, L.E., Zottola R., Asp, P.A. and Roet, H.D. (1988) 'Structural systems for the new Bank of China Building, Hong Kong', *Proc. 4th Int. Conf. on Tall Buildings*, Hong Kong University Press.

Rosenblueth, E. (1973a) 'Code specification of safety and serviceability', *Planning and Design of Tall Buildings, Proc. 1972 ASCE-IABSE Int. Conf.*, Vol. 1b, ASCE, New York, pp. 931–960.

Rosenblueth, E. (1973b) 'Floor loads', *Planning and Design of Tall Buildings, Proc. 1972 ASCE-IABSE Int. Conf.*, Vol. 1b, ASCE, New York, No. 5-D1.

Rosenblueth, E. (1974) 'Analysis of risk', *Proc. 5th World Conf. on Earthquake Engineering*, Rome, Vol. 2, pp. CIL-CLVIII.

Schwerdt, R.W., Ho, F.P. and Watkins, R.R. (1979) *Meteorological Criteria for Standard Project Hurricane and Probable Maximum Hurricane Wind Fields, Gulf and East Coasts of the United States*, NOAA Tech. Rep. NWS 23, Washington, DC.

Seaton, B. (1976) 'The detection of periodicity in irregular data', *J. Theor. Biol.*, **63**, 311–324.

Seed, H.B. and Idriss, I.M. (1969) 'Influence of soil conditions on ground motions during earthquakes', *J. Soil Mech. Found. Div., ASCE*, **95**, no. SM1, Proc. Paper 6347, pp. 99–137.

Seite, F., Jastron, R. and Nierenburg, W. (1989) *Scientific Perspectives on the Greenhouse Problem*, Marshall Institute, Washington, DC.

Severn, R.T., Jeary, A.P. and Ellis, B.R. (1980) 'Forced vibration tests and theoretical studies on dams', *Proc. Inst. Civ. Eng., Part 2*, **69**, 605–634; and discussion (1981) *Proc. Inst. Civ. Eng., Part 2*, **71**, 575–595.

Shears, M. (1970) 'Problems in the application of statistical design methods', *The Modern Design of Wind Sensitive Structures, Proc. CIRIA Seminar*, London.

Shellard, H.C. (1963) 'The estimation of design wind speeds', *Proc. Conf.*, NPL, Teddington, Vol. 1, pp. 29–52.

Shinners, S. (1964) *Control System Design*, Wiley, New York.

Shiotani, M. and Iwatani, Y. (1979) 'Gust factors over flat terrains and their modification by a barrier', *5th Int. Conf. on Wind Engineering*, Colorado, pp. 203–214.

Simiu, E. (1975) 'Equivalent static wind loads for tall building design', *Proc. 4th Int. Conf. on Wind Effects on Buildings and Structures*, London, Cambridge University Press.

Simiu, E. and Lozier, D.W. (1975) *The Buffeting of Tall Structures by Strong Winds*, Building Science Series 74, NBS, Gaithersburg, MD.

Smeaton, J. (1759) 'An experimental investigation concerning the natural powers of water and wind', *Phil. Trans. R. Soc.*, **51**, 1759–1760.

Snaebjornsson, J.T. and Sigbjornsson, R. (1992) 'Wind structure over postglacial lava surface', *J. Wind Eng. Ind. Aerodyn.*, **41–44**, 305–315.

Soroka, W.W. (1949) 'Note on the relations between viscous and structural damping coefficients', *J. Aeronaut. Sci.*, pp. 408–409 and 448, July.

Sparks, P.R., Reid, G.T., Reid, W.D., Welsh, S. and Welsh, N. (1992) 'Wind conditions in Hurricane Hugo by measurement, inference, and experience', *J. Wind Eng. Ind. Aerodyn.*, **41–44**, 55–66.

Steffens, R.J. (1973) *Structural Vibration and Damage*, Building Research Establishment, Garston, UK.

Stubbings, B.J. (1990) 'Appraisal of the structural adequacy of high rise reinforced concrete domestic buildings in Hong Kong', *Struct. Eng.*, **68**, no. 16, 317–326.

Takeuchi, M. (1960) 'Vibrational characteristics of buildings – Part 1, Vibrational characteristics of actual buildings determined from vibration tests', *Proc. 2nd Conf. on Earthquake Engineering*, Tokyo.

Tamura, Y. (1990) 'Suppression of wind-induced vibrations of buildings', *J. Wind Eng.*, no. 44, July.

Tamura, Y., Sasaki, A. and Tsukagoshi, H. (1991) 'Evaluation of damping ratio of randomly excited structures', *J. Wind Eng.*, no. 47, April.

Tamura, Y., Yamada, M. and Yokota, H. (1994) 'Estimation of structural damping of buildings', *ASCE Structures Congr. and the IASS Int. Symp.*, Atlanta, GA.

Taylor, G.I. (1920) 'The phenomena of rupture and flow in solids', *Proc. R. Soc.*, 21 October, pp. 162–198.

Taylor, P.A., Mickle, R.E., Salmon, J.R. and Teunissen, H.W. (1983) *The Kettles Hill Experiment: Site Description and Mean Flow Results*, Report AQRB-83-001-L, Atmospheric Environment Service, Downsview, Ontario.

Teunissen, H.W. (1983) 'Wind-tunnel and full-scale comparisons of mean wind flow over an isolated low hill', *6th Int. Conf. on Wind Engineering*, Gold Coast, Australia.

Tielman, H.W. (1992) 'Wind characteristics in the surface layer over heterogeneous terrain', *J. Wind Eng. Ind. Aerodyn.*, **41–44**, 329–340.

Tielman, H.W. and Mullins, S.E. (1979) 'The structure of moderately strong winds at a mid-Atlantic coastal site (below 75 m)', *5th Int. Conf. Wind Engineering*, Fort Collins, Colorado, pp. 145–160.

Tsai, I.C., Lin, T.W. and Chiu, A.N.L. (1984) 'Full-scale experimental and analytical study of dynamic response of free-standing lattice tower to typhoons', *Proc. CCNAA–AIT Joint Seminar on Research for Multiple Hazards Mitigation*, Tainan, Taiwan.

Twisdale, L.A. and Vickery, P.J. (1992) 'Research on thunderstorm wind design parameters', *7th US Nat. Conf. on Wind Engineering*, Vol. II, pp. 823–832 and *Int. J. Wind Eng. Ind. Aerodyn.*

Van der Hoven, I. (1957) 'Power spectrum of horizontal wind speed in the frequency range from 0.0007 to 900 cycles per hour', *J. Meteorol.*, **14**, 160–164.

Van Koten, H. (1977) 'Structural damping', *Heron*, **22**, no. 4.

Vickery, B.J. (1966) 'On the assessment on wind effects on elastic structure', *Civ. Eng. Trans., Inst. Eng., Aust.*, pp. 183–192.

Vickery, B.J. (1969) *On the Reliability of Gust Loading Factors*, NBS Building Science Series 30, US Department of Commerce, Washington, DC.

Vickery, B.J. (1972) 'On the aeroelastic modelling of structures in wind', *Proc. Conf. on Structural Models*, Sydney, Cement and Concrete Association of Australia in conjunction with Department of Architectural Science, University of Sydney, and Institution of Engineers, Australia, NSW Division.

Vickery, B.J. (1973) 'Notes on wind forces on tall buildings', *Proc. Conf. on Planning and Design of Tall Buildings*, Sydney University.

Vickery, B.J. (1976) 'The design and performance of a low-cost boundary layer wind tunnel', *Proc. 1974 2nd USA/Japan Research Seminar on Wind Effects on Structures*, University of Tokyo Press.

Vickery, B.J. and Apperley, L.W. (1973) *On the Prediction of the Ground Level Wind Environment*, Civil Engineering Rep. R227, University of Sydney.

Vickery, B.J. and Davenport, A.G. (1967) 'A comparison of theoretical and experimental determination of the response of elastic structures to turbulent flow', *Proc. Int. Conf. on Wind Effects on Buildings and Structures*, Ottawa, University of Toronto Press.

Vogt, S. and Thomas, P. (1993) 'SODAR: a useful remote sounder to measure wind and turbulence', *3rd Asia–Pacific Symp. on Wind Engineering*, pp. 435–440.

Von Karman, T. (1948) *Proc. Natl Acad. Sci. USA*, **34**, 530.

Walker, G.R. (1975) *Report on Cyclone Tracy – Effects on Buildings – December 1974*, Vol. 3, Australian Department of Housing and Construction.

Wieringa, J. (1992) *Updating the Davenport Roughness Classifications*, Elsevier, Amsterdam.

Wyatt, T.A. (1977) 'Mechanisms of damping', *Symp. on Dynamic Behaviour of Bridges*, Transport and Road Research Laboratory, Crowthorne, Berkshire, UK.

Yeomans, D.T. and Cottam, D. (1989) 'An architect/engineer collaboration: the Tecton/Arup flats, *Struct. Eng.*, **67**, no. 10, and discussion **67**, no. 19.

Standards and design guides

Architectural Institute of Japan, *AIJ Recommendations for Loads on Buildings*, Tokyo, 1996.

ASCE Committee on Wind Effects, *Wind Loading and Wind Induced Structural Response*, ASCE, 1987.

ASCE Minimum Design Loads for Buildings and Other Structures (Revision of A58.1), ASCE 7–88, 1990, New York

BRE Digest 346, Parts 1 to 8, *The Assessment of Wind Loads. 1988–1990*, Building Research Establishment, Garston, UK.

BSI, *BS 6399 Part 2. Code of Practice for Wind Loads.* Building and Civil Engineering Sector Board, British Standards Institution, London, 1995.

Building Development Dept., *Code of Practice on Wind Effects, Hong Kong – 1983.*

Bureau of Mines Report of Investigations, RI 8507. *Structure Response and Damage Produced by Ground Vibration from Surface Mine Blasting*, United States Department of the Interior, 1980.

CEN–European Committee for Standardisation, *Eurocode 8. Common Unified Rules for Structures in Seismic Regions*, ENV, 1998-1-1.

Code SIA 160 (Swiss Code), *Actions on Structures*, Schweizer Ingenieur- und Architekten-Verein (SIA), Zurich, 1989.

Cook, J., *The Designer's Guide to Wind Loading of Building Structures*, BRE, 1989.

DIN 4150, Protection against Vibration in Building Construction, German Institute for Standards, Berlin, (1939).

Engineering Sciences Data Unit: Item 76001, *The Treatment of Flexible Structures*, ESDU, London, 1976.

Engineering Sciences Data Unit: Item 82026, *Strong Winds in the Atmospheric Boundary Layer: Part 1 Mean hourly wind speeds; Part 2 Discrete gust speeds*, ESDU International, London, 1976.

Jeary, A.P., *Damping in Tall Buildings*, Conference on the Second Century of the Skyscraper, ASCE/Council on Tall Buildings, Chicago, 1986.

Jeary, A.P. and Sparks, P.R., *Some Observations on the Dynamic Sway Characteristics of Tall Buildings*, BRE Current Paper CP7/78, 1978.

Jeary, A.P., *Damping in Chimneys*, CICIND Conference, Hong Kong, 1987.

Jeary, A.P., *In-service Performance of Bridges*, China Civil Engineering Society Congress, Shanghai, 1986.

Jeary, A.P. and Irwin, A.W., Vibrations of a nuclear power station charge hall, *Int. J. Earthq. Eng. Struct. Dyn.*, **4**, 221–229, 1976.

Jeary, A.P., Ellis, B.R. and Vasconcellos, R., *The Practical Vibration Analysis of Structures with Particular Reference to Offshore Structures*, Conference Brasil Offshore, Session 6 COPPE, UFRJ, 1979.

Jeary, A.P. and Irwin, A.W., *Vibrations of Power Station Buildings*, CERL/RD/L/N276/73, 1973.

Jeary, A.P., Beak, M. and Littler, J.D., *Dynamic Testing of the Dance Floor at Middlesbrough Town Hall*, March 1982.

Jeary, A.P., *The Impact of Modulated Feedback on Design Criteria in a Non-linear Environment*, Tall Buildings 2000 and Beyond. Also presentation of work to conference held in Hong Kong 5th to 9th November 1990.

National Council of Canada, *Supplement to the National Building Code of Canada, 1990*, NRCC No. 17724, Ottawa, 1991.

National Standard of the People's Republic, *Load Code for the design of building structures GBJ 11–87*, Department of Standards and Norms, Ministry of Construction; English translation by New World Press, Beijing, 1994.

NCEER Draft New York Seismic Code, National Center for Earthquake Engineering Research, Buffalo, USA, 1991.

Ostrowski, J., Jeary, A.P. and Grace, W.R., *The Role of Vibration Testing for Assessment of a Damaged Wharf*, International Conference on Case Histories in Structural Failures, Singapore, 1989.

SEAOC *Recommended Lateral force requirement and tentative commentary*, Seismology Committee, Structural Engineers Association of California, USA.

Standards Australia, *Minimum Design Loads on Structures* (known as the SAA Loading Code), Part 2, Wind Loads, AS 1170.2–1989.

UK Code of Practice on Wind Loading, CP3 Chapter 5 Part 2, *Wind Loading*, British Standards Institute, London, 1972.

Index

Page numbers in **bold** refer to Figures